Research & Education

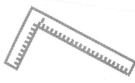

The Handyman's
Practical
Guide to
Renovating
Houses
for Increasing Value and Profit

Gerald E. Sherwood, USDA Forest Service

In cooperation with the

National Association of Home Builders

Research & Education Association
61 Ethel Road West

The Handyman's Practical Guide to Renovating
Houses for Increasing Value and Profit

Printed in the United States of America

Library of Congress Control Number 2001098089

International Standard Book Number 0-87891-425-0

Research & Education Association
61 Ethel Road West
Piscataway, New Jersey 08854

What This Book Will Do For You

With the prices of new homes and real estate skyrocketing across the United States, both current and prospective homeowners are looking to renovation as a way to add value to their homes, save money, and create living spaces that meet their needs and reflect their individual style.

Renovating a less expensive older house can cost considerably less than purchasing a newer house with a higher price tag, and sellers of renovated homes often find that they can recoup several times their investment.

Depending on the economic climate and location of the houses, both buyers and sellers can reap large personal and financial benefits from either renovating a house on their own, or engaging professional contractors to do the work.

Whether you are already living in the home to be renovated, or you are contemplating the purchase of a "handyman's special" this book can be an invaluable resource.

About Research & Education Association

Research & Education Association (REA) is an organization of educators, scientists, and engineers specializing in various academic fields. Founded in 1959 with the purpose of disseminating the most recently developed scientific information to groups in industry, government, and universities, REA has since become a successful and highly respected publisher of study aids, test preps, handbooks, and reference works.

REA's publications and educational materials are highly regarded and continually receive an unprecedented amount of praise from professionals, instructors, librarians, parents, and students. Our authors are as diverse as the fields represented in the books we publish. They are well-known in their respective disciplines and serve on the faculties of prestigious high schools, colleges, and universities throughout the United States and Canada.

Dr. M. Fogiel, Director
Carl Fuchs, Chief Editor

Contents

Introduction   **1**

Purposes of this handbook   1

Units of measure   2

Basic considerations   2

Special considerations   2

     Health aspects   3

     Fire safety   3

     Energy efficiency   3

Planning   4

     Establishing objectives and priorities   4

     Budgeting for rehabilitation   5

     Regulations, codes, and planned development   7

Condition Assessment   **9**

Initial considerations   9

     Professional assistance   9

     Choosing and working with professional contractors   9

Complying with codes and regulations   10

Financing projects   10

     Liens   10

     Arbitration   11

Health and safety assessment   11

     Radon   11

     Lead   13

     Asbestos   16

     Fire safety   18

     Security   20

Thermal assessment and moisture control   20

     Climatic and site considerations   20

     Heat transfer   20

     Types of insulation   21

     Vapor retarders   22

     Assessment of insulation   24

     Airtight house and foundation   26

     Symptoms of excess moisture   26

The site   27

     Grading   27

     Landscaping   28

     Walkways, driveways, and patios   28

     Fencing   29

Interior layout   29

     Preservation and restoration   29

     Accessibility for special groups   30

     Entryway   30

     Living rooms, family rooms, and dining areas   30

     Bedrooms   30

     Kitchens and bathrooms   30

     Expansion of available space   31

     Layout change of more recent homes   31

Foundation 31
    Normal settlement 32
    Soil moisture and foundation settlement 36
    Checking foundations 36
Exterior walls 37
    Siding 37
    Windows 39
    Doors 40
    Exterior finishes 40
Roofing 42
    Wood shingles 42
    Asphalt shingles 43
    Built-up roofing 44
    Other types of roofs 44
    Flashing 44
    Roof overhang 44
    Soffit and fascia 45
Chimneys and fireplaces 45
Exterior utilities 47
Porches and decks 47
Basement 47
    Moisture 47
    Other problems 48
Living areas 49
    Wood floors 49
    Tile and sheet flooring 49
    Walls and ceilings 49
    Interior coatings 50
    Stairs 50
Attic 51
Kitchen 52
    Layouts 52
    Appliances 54
    Floors 54
Bathrooms 55
    Fixtures 58
    Floors 58
Utility assessment 58
    Heating 59
    Air-conditioning 59
    Water system 60
    Electrical system 61
Recognition of damage 62
    Decay 62
    Insects 69
    Rodents 74

Rehabilitation **77**
Introduction 77
Hazard control 77
    Radon 77
    Lead 81
    Asbestos 85
    Fire safety 89

Thermal protection and moisture control 93
    Installment of insulation 93
    Vapor retarders 96
    Indoor humidity control 99
    Site improvements 99
Major structural features 101
    Foundations and basements 101
    Floor systems 108
    Wall systems 121
    Wall openings 124
    Roof systems 137
    Repair of structural decay and insect and rodent damage 146
Siding 147
    Replacement techniques 148
    Panel siding 149
    Horizontal wood siding 152
    Vertical wood siding 154
    Wood shingle and shake siding 156
    Aluminum and vinyl siding 158
    Stucco finish 158
    Masonry veneer 158
    Material transition 158
Exterior finishes 159
    Types of exterior wood finishes 159
    Application 162
    Refinishing 166
    Correction of finishing problems 169
Interior structural changes 176
    Floor plan 177
    Basement conversion 182
Interior finishing 182
    Types of finishes 182
    Wood floors 185
    Removal of existing finishes 186
    Safety considerations 186
Vertical expansion 187
    Attic space 187
    Shed dormer 188
    Gable dormer 188
    Other considerations 189
Horizontal expansion 191
Kitchen remodeling 191
Heating, air-conditioning, plumbing, and electrical systems 192
    Cutting floor joists 192
    Utility walls 193
Fireplaces, woodstoves, and chimneys 193
    Conventional masonry fireplaces 193
    Air-circulative fireplace forms 195
    Prefabricated fireplace units 196
    Woodstoves 196
    Masonry chimneys 198
    Insulated steel chimneys 199

Garages and carports 200
    Addition of garages or carports 200
    Garage doors 200
    Carports 201
    Conversion of garage to living space 201
Deck, patio, and porch additions 203
    Planning and design 203
    Material selection 204
    Structural considerations 204
    Fasteners 207
    Finishes 207

Maintenance **209**
    General considerations 209
    Exterior wood finishing 209
        Walls 209
        Decks and porches 210
        Maintenance of wood roofs 210

Caution **211**

Appendix **212**

Glossary **217**

Introduction

From the earliest days of settlement in America until the present time, wood has been a preferred material for home construction. Its popularity reflects its easy availability, moderate cost, attractiveness, high suitability for numerous uses in home construction, and proven durability. A well-built and properly maintained wood house will retain its strength and structural properties for hundreds of years.

If not cared for, wood houses may deteriorate. Long neglect can greatly increase the cost of rehabilitation. Unchecked deterioration can reach the point at which rehabilitation is impractical.

America possesses a large stock of well-built older homes for which rehabilitation makes good structural and economic sense. Eliminating their problems and defects can create the functional equivalent of a new home, often at significant savings over the cost of a new home of comparable quality. In addition, such rehabilitation makes a valuable contribution to the preservation of communities and neighborhoods and contributes to the upgrading of one of the Nation's largest and most important resources—its stock of available family dwellings.

When an older home is being rehabilitated, the owner may also want to change or modernize certain features. Such changes are not necessary to maintain or protect the structure but are made to reflect the owner's tastes, to accommodate changes in styles that have occurred since the home was built, and to provide enhancements for more comfortable living.

Americans devote substantial time, effort, and resources to the rehabilitation and improvement of their homes. In recent years Americans spent more than $100 billion on home rehabilitation and remodeling.

Purposes of This Handbook

This handbook provides detailed information for assessing the condition of older homes and for rehabilitating and maintaining them. It also provides information on changes and enhancements that can be made as part of the rehabilitation process. In the "Condition Assessment" section of the book, we assume the reader does not necessarily own the home. In the "Rehabilitation" section, we assume the reader does own the home.

The section on in-place evaluation is a guide for determining the suitability of the wood-frame structure for rehabilitation. We present a systematic approach for inspecting the building and evaluating information from the inspection. If the structure is deemed worthy, the rehabilitation portion is a guide for planning and conducting the work.

The book gives some general guidelines for assessing the condition of heating, plumbing, and electrical systems and some general information on their rehabilitation or replacement. However, professional assistance is likely to be desirable in these areas.

This handbook should be particularly useful for owners and prospective buyers of older homes. It should also be of interest to carpenters, contractors, and lending institutions, as well as groups that seek to maintain and improve homes within a community.

This handbook is designed as a reference for members of the building trades and professions. It is equally intended to serve the needs of persons with skills in home rehabilitation and repair and as a reliable source of information and guidance for home dwellers and members of the general public who may not undertake the actual work of rehabilitation themselves but who wish to understand what needs to be done and how it is done.

Keeping in mind the various nonprofessional audiences, the authors presume no previous specialized knowledge of home construction or the building trades.

Units of Measure

Mostly English units are used in this book. Table 1 provides conversion factors to equivalent metric units for all English units used in this book.

Table 1—Conversion factors for units in this book

English unit	Conversion	Metric unit
in.	x 25.4	mm
ft	x 0.3048	m
ft^2	x 0.09290304	m^2
ft^3	x 0.02831685	m^3
°F	$t_c = (t_F - 32)/1.8$	°C
mils	x 0.0254	mm
lb/in^2	x 6894.757	Pa
lb/in^3	x 27679.9	kg/m^3
lb/ft^2	x 47.88026	Pa
gal	x 3.785412 x 10^{-3}	m^3
gal/min	x 6.309020 x 10^{-5}	m^3/s
perm	x 5.72135 x 10^{-11}	kg/Pa · s · m^2
pCi/L (picocuries per liter)	x 0.037	Bq/L

Basic Considerations

It goes without saying that certain basic considerations may determine the feasibility of rehabilitating an existing house. For example, if some of the main structural elements (girders, joists, studs) have become inadequate because of decay or other damage, or were initially inadequate for modern requirements (too small or too low in grade), such deficiencies must be considered thoroughly and early in the planning process. Perhaps roof or wall coverings have deteriorated to a dangerous point, or uneven settlement has caused the house to rack badly. These, too, must be considered early in the process. Such deficiencies may, in fact, dictate that the rehabilitation be completely abandoned because of inordinate costs.

Special Considerations

Three special considerations relate to a number of features of home rehabilitation. They are

- health aspects of the indoor environment,

- fire safety, and

- energy efficiency.

Health Aspects	A number of health matters must be considered when rehabilitating an older home. Prominent among these are the possible presence of radon, asbestos, and lead.

Radon

Radon is a colorless, odorless, naturally occurring radioactive gas that is present virtually everywhere in the earth. It can enter houses from the ground, where it disperses less readily than it does outdoors. Radon decays into unstable elements that can damage lung cells and cause lung cancer. The Environmental Protection Agency (EPA) estimates that radon causes between 5,000 and 20,000 deaths from lung cancer each year.

Asbestos

Asbestos is a silica-based mineral that is found in various building materials. It has been used in such homebuilding applications as thermal insulation of pipes and in vinyl-asbestos floor tile. Asbestos causes no harm when undisturbed, but when asbestos insulation ruptures or vinyl-asbestos tile breaks up, which could occur during rehabilitation, invisible asbestos fibers are released into the air. If these fibers are inhaled, they can cause asbestosis, a degenerative lung disease, and lung cancer.

Lead

Lead was used for water piping in older homes and as a constituent in solder used to join copper piping. Lead may also be present in older house paints; such paints are found in most houses built prior to 1940 and in many that were built later. Lead poisoning can be caused by lead in drinking water or even by breathing the dust in houses that have interior lead paint. Minimal levels of lead in the blood of children can have serious health effects, including partial loss of hearing, retardation, inhibited metabolism of vitamin D, and disturbances in blood formation.

Fire Safety

Every year in the United States, there are more than 600,000 residential fires serious enough to be reported. The ensuing property losses exceed $7 billion annually. These fires account for 80 percent of fire deaths, 65 percent of injuries, and 46 percent of structural property loss in the United States. Despite progress in reducing these losses, the United States continues to have one of the highest fire death and injury rates in the world. A disproportionate number of victims have been people who are unable to help themselves—children, the elderly, and those unable to escape on their own.

These data indicate strongly that fire safety must be considered in the assessment of and planning for the rehabilitation of a house, regardless of the objective of the rehabilitation. Among the factors that should be considered are the availability and convenience of means for emergency egress in case of fire, particularly for disabled and elderly persons; the availability and condition of smoke and fire detectors; the feasibility of installing a sprinkler system; the condition of protection around wood-burning stoves; the capacity and condition of the internal electric distribution system; and the condition of the central heating system and its adjuncts such as chimney flues. In some jurisdictions, local ordinances may dictate that such things as smoke detectors must be installed in new houses. Smoke detectors may be required in rehabilitated houses.

Energy Efficiency

The recent increased concern for energy efficiency is reflected in many features and aspects of new home construction. Rehabilitating an older home is an opportunity to incorporate features—not likely to be found in such homes—that promote energy conservation and the associated lower operating cost. As a major aspect of renovation, the owner should take an integrated approach to the house's energy efficiency, including the house envelope and heating, ventilating, and air conditioning systems, rather than considering such aspects independently.

Planning

Planning a rehabilitation project involves

- establishing objectives and priorities;

- budgeting and financial considerations, including the probability of recovering costs if and when the house is sold; and

- determining the applicability and impact of code requirements and restrictions.

Establishing Objectives and Priorities

A basic step in planning for rehabilitation of an older building is to establish the overall objectives. Different objectives call for different approaches, some of which are discussed here.

Once the objective of a proposed rehabilitation project has been established, the steps considered necessary or desirable to reach this objective can be set up and their priorities assigned. It seems obvious that the basic considerations and the special considerations discussed earlier must be assigned the higher priority because they relate to the safety of the home and its occupants. Other considerations, such as additions to the house or modernization of kitchens or bathrooms, are not as important. In many cases, economics may dictate whether or not the low-priority steps are feasible.

Creation of a Sound, Livable Home

If this is the basic objective, the owners can proceed to a second level of considerations: namely, to differentiate steps that are required to protect the basic integrity and value of the structure from those that are designed to reflect the family's tastes and to make it more comfortable and livable. If it is economically infeasible or unwise to invest in all the desired extras, the basics must come first, and the extras must be ranked in order of priority.

Personal or Sentimental Considerations

Previous ownership of a building by members of one's family or a desire to remain in a given area or neighborhood can justify a somewhat greater investment in rehabilitation than that dictated by pure economics.

Historic Preservation

Historic preservation is a broad concept that encompasses the restoration of features of a structure, or the entire structure, to its original appearance. This type of project generally involves specialized knowledge, materials, and workmanship, and is usually more expensive than standard rehabilitation.

The subject of historic preservation lies largely beyond the scope of this book. Persons interested in historic preservation as it pertains to their home will find an abundance of resources to assist them. As public interest in historic preservation has increased, the field has generated substantial literature, a number of practicing specialists, and at least one periodical, *The Old House Journal* which covers subjects relating to homes built before 1939.
Information on historic preservation is also available from the National Park Service.

A few comments, however, are pertinent to the rehabilitation of most homes. First, every older house is historic in the sense that houses are major artifacts of their time. They reflect architectural, social, cultural, and aesthetic features of the period when they were built. Second, changes in older homes are often desirable and appropriate to make them more livable today, but the homeowner will also wish to approach such changes in a way that creates aesthetic harmony with the original house and neighborhood. As part of a rehabilitation plan, the owner may wish to include preservation or restoration of distinctive or attractive features of the original dwelling.

Rehabilitation for Resale

An important question with regard to investment in rehabilitation is the length of time that the family plans or expects to live in the house if, in fact, they plan to live in it at all. In some instances, the owner of an older house may be contemplating resale, either immediately after completion of rehabilitation or following a certain period of residence after the work has been finished. This can affect both how much money is spent and what is done.

When rehabilitating a house for investment purposes, one is, in effect, fixing it up for someone else to live in. Investment priorities are likely to be influenced by the desire to make the house as attractive as possible to the largest number of potential buyers while reducing expenditures to a minimum.

When rehabilitating an investment property, the owner should pay particular attention to the average value of houses in the neighborhood, which is likely to have a decisive impact on the maximum sale value of the home that is being rehabilitated. This matter is discussed in following sections.

In a neighborhood containing a number of somewhat similar older homes, an attractive rehabilitation can significantly enhance the salability of a home compared with that of its neighbors.

Particular attention should be given to the return that one might expect from different types of rehabilitation and improvement projects. For example, a major kitchen renovation may be highly desirable if the family plans to live in the home. Large expenditures for extensive kitchen remodeling, however, generally offer a smaller return on investment for resale than minor remodeling to deal with basic inadequacies. Additional information on the return an owner can expect from various types of rehabilitation projects is provided in Table 2.

If the owner plans to reside in the house before selling it, the cost of any changes to enhance his or her stay should be carefully considered.

Budgeting for Rehabilitation

Budgeting for rehabilitation should consider

- average value of homes in the neighborhood,

- budgeting for continuing maintenance, and

- return on the investment that is made in rehabilitation.

Table 2—Predicted percentage of investment recovery on rehabilitation projects

Type of job	Investment recovery (percent)	
	Done professionally	Materials alone
Interior facelift	90	173
Bathroom	74	205
Kitchen	73	160
Fireplace addition	68	159
Exterior painting	60	1,109
Wood siding	38	100
Window and door replacement	38	121
Vinyl siding	31	193

The average value of homes in the neighborhood of the house to be rehabilitated has an important impact on the likelihood of recovering the investment in rehabilitation. Frequently, owners will spend a substantial amount of money to rehabilitate and improve basic structural matters, and to incorporate long-desired changes and enhancements. Then, when the house is sold, the owners may find that the sale price covers only a very small portion of their investment.

Such cases strongly support a rule of thumb that is widely used in the real estate industry: a house will rarely sell for more than 15 percent over the median price of other homes in the neighborhood, regardless of special features or the extent of rehabilitation. This, in turn, reflects the fact that a large proportion of all American homes are located in neighborhoods where home values fall within about 10 percent of the median price in that neighborhood.

Certain types of neighborhoods offer exceptions to this rule, particularly city neighborhoods where homes of widely varying value may be located within a few blocks of each other. In most suburban neighborhoods, however, ignoring the 15-percent rule is likely to take its toll on the investment that is made.

Real estate agents can provide information on current and anticipated neighborhood values and the price that one is likely to secure for one's own house. This information must be considered in setting objectives and priorities.

*Budgeting for
Maintenance*

A common oversight in budgeting for rehabilitation is failing to provide for continuing maintenance, which is a foreseeable expense.

The following list provides an approximate range of life expectancy for certain home equipment, systems, features, and improvements. Although it does not include everything and it does not predict each individual improvement, it shows both the inevitability and the significant cost of continuing home maintenance.

Furnace	20–30 years
Electric baseboard heaters	15 years
Heat exchanger	5–30 years
Air-conditioning compressor	10–15 years
Water heater, gas or electric	8–12 years
Washing machine	8–14 years
Dryer	10–20 years
Kitchen wall oven	15–20 years
Roofing, asphalt shingles	15–30 years
Roofing, wooden shingles	10–20 years
Interior paint	5 years

Return on Investment

Studies and surveys consistently show that, within the general limits posed by the 15-percent rule previously described, rehabilitation of homes increases their resale value. The homeowner can therefore enjoy the benefits of living in the rehabilitated home and recover at least part of the investment when he or she sells the property. In some instances, the homeowner may realize a net profit.

In its May/June 1987 issue, the magazine *Practical Homeowner* reported the results of a detailed study conducted with the cooperation of a panel of 14 professional appraisers from 14 metropolitan areas throughout the country. The appraisers were asked to estimate the market value in their area of a hypothetical 25-year-old, three-bedroom ranch house with 1-1/2 baths and a full basement, situated on 1/4-acre lot in an average, middle-class neighborhood. The appraisers were then given a list of potential rehabilitation projects for this hypothetical home, with a description of the scope of each project. They were asked to provide estimates as follows:

- The cost of materials only (basically, the cost of the project if the homeowner performed the work)

- The combined cost of materials and labor if the work was done professionally

- The amount that the rehabilitation/improvement would add to the market value of the home

From these figures, the magazine calculated the likely percentage of investment recovery that a homeowner could expect if he or she did the work and the likely percentage of recovery if the work was done by professionals. Results for a number of projects are shown in Table 2.

Of course, these figures can vary significantly in individual cases and geographic locations.

Regulations, Codes, and Planned Development

Local zoning regulations or building codes, or both, may affect various features of a rehabilitation project. It is good practice to discuss the planned rehabilitation with local offices that administer zoning regulations and building codes before doing any work. When meeting with these officials, a written summary of the proposed project, with sketches, may be helpful.

Before deciding on the extent of the rehabilitation, it is also good practice to become familiar with area developments that might affect property values positively or negatively. These could include proximity to existing or planned roadways, public transportation, schools, shopping centers, libraries, or community centers, and the foreseeable course of commercial and residential development in the area.

In some localities, codes specify that if rehabilitation involves more than half of the existing structure, requirements for new construction may apply. If major rehabilitation is planned, the question of which code requirements are applicable should be discussed with local officials.

Condition Assessment

This portion of the book is designed to help the reader determine

- the general condition of the house,

- specific steps that are required to maintain its basic structural integrity, and

- the feasibility and suitability of various enhancements to make the house more comfortable, livable, and attractive.

Initial Considerations

Professional Assistance

A number of professional firms and individuals offer services for assessing the condition of homes and for estimating the cost of rehabilitation. Many of these firms and individuals are exclusively in the business of home condition assessment, and are often retained by potential buyers. Their services are also available to homeowners who are already living in the home to be examined.

Choosing and Working With Professional Contractors

The Council of Better Business Bureaus, the National Association of Consumer Agency Administrators, the U.S. Office of Consumer Affairs, and the Remodelers™ Council of the National Association of Home Builders (NAHB) have worked together to create a set of guidelines to assist homeowners in choosing and working with a reputable contractor for home remodeling and rehabilitation. The information and guidance provided by these groups appeared initially in a jointly issued publication entitled *Choosing a Professional Remodeling Contractor*. A revised edition, published by the NAHB Remodelers™ Council, is entitled *How to Choose a Remodeler Who's on the Level*. Some highlights from the publication follow.

- Solicit two or three bids for the work you are considering. Make sure all bids are based on the same set of specifications. Do not automatically accept the lowest. Discuss the bids in detail with the contractors.

- Ask for local references, and call them to see if they were satisfied with the contractor's work.

- Check local sources such as the Better Business Bureau and government consumer affairs office to see if they have information about the contractor or contractors you are considering. Ask how long the company has been in business, and find out if there are complaints or other relevant information about the firm on file.

- Check to see if the contractor is a member of a professional association that has a code of ethics or standards for remodelers. Find out if members have pledged to arbitrate disputes.

- Ask the contractor if the company has insurance that covers workers' compensation, property damage, and personal liability in case of accidents. Ask to see a copy of the certificate of insurance. In some areas, such insurance is required by law.

- If licensing and/or bonding are required in your area, ask to see a copy of the license or bond.

- Get all oral statements and agreements in a written contract. If you intend to do some of the work yourself or to hire another contractor to do it, this should appear in the contract. Read and understand everything before you sign. Among other things, the contract should specify all materials to be used, approximate starting and completion dates, cleanup and removal of debris, special requests such as saving lumber for firewood or saving certain materials or appliances, and areas where materials may not be stored. Be sure the financial terms are clear. The contract should include the total price, the schedule of payments, and whether or not there is a cancellation penalty.

- If a warranty is offered, get it in writing, read it carefully, and be sure you understand its terms and conditions.

- You should expect to make a down payment of about one-third of the total contract price when signing. Except for this payment, do not make payments for work that has not been completed. Schedule additional payments at weekly or monthly intervals or after completion of each phase of the project.

- It is recommended that you withhold a negotiated percentage of the contract price, typically 10 percent, until the job is completed. If a building permit was required, do not release the remaining money until the building inspector has approved the job.

- Never sign a completion certificate until all work called for in the contract has been completed to the owner's satisfaction. Lenders usually require a signed completion certificate before they will release the last payment.

Complying With Codes and Regulations

Various features of your proposed rehabilitation may relate to local zoning regulations. Zoning requirements should be checked, and any needed permits or permissions should be obtained prior to commencement of work.

Most communities and jurisdictions have adopted building codes. These codes vary in their requirements, but they generally specify that a building permit is required whenever structural work is involved or when the basic living area of a home is to be changed. The contractor should obtain the necessary permits.

Planned communities and areas designated as historic districts may have architectural review boards that must approve all rehabilitation that will be visible on the exterior. For both planned communities and historic districts, the requirements generally include visible structural work, exterior refinishing, and all changes in windows and doors.

When a building permit has been issued, the local issuing agency will inspect the work when it has reached a certain stage or when it is completed, or both, to ensure compliance with code requirements. It is the contractor's responsibility to arrange these inspections.

Financing Projects

Various sources of financing are available, including personal loans, home equity loans, credit unions, insurance policies, banks, and savings and loan institutions. Compare features such as interest rates, terms, and tax considerations from the different sources.

Liens

For large rehabilitation projects, protect yourself from liens against your home in the event the contractor does not pay subcontractors. Local laws about liens vary, so a lawyer's assistance may be desirable to secure the requisite protection.

| Arbitration | The Better Business Bureau and the National Association of the Remodeling Industry conduct a program called Remodelcare™. Participating contractors agree to |

- subscribe to the Remodelcare ™ code of ethical conduct,

- adhere to the Better Business Bureau standards of practice for the home improvement industry, and

- sign a pledge to arbitrate disputes that cannot be resolved through mediation.

Information on the program can be obtained from local Better Business Bureaus or a comparable governmental agency. If the contractor you are considering does not participate in this program, you may wish to have an arbitration clause written into your contract.

Health and Safety Assessment

You should consider several health and safety factors when assessing a home for rehabilitation. These factors, and the means for assessing them, are described here. Their mitigation is discussed under "Hazard Control" in the "Rehabilitation" section of this book.

Radon

Radon is an invisible gas that is a byproduct of the radioactive decay of uranium. It occurs everywhere in the earth's crust, although concentrations vary in different locales. In addition to naturally occurring concentrations, factors such as soil type and porosity affect the amount of radon that migrates toward and into a home.

The level of radon in the atmosphere is too low to pose significant health problems. However, radon can enter the house through cracks in basements and foundation walls, sumps, and drains. In the enclosed space of the house, it can reach concentrations that exceed the levels in the general atmosphere, sometimes reaching hazardous levels. Radon decays into several short-lived elements called radon daughters. These elements—polonium 218, lead 214, bismuth 214, and polonium 214—which are solids rather than gases, undergo further decay in a series of processes, that ultimately produce a stable element—lead 206.

The decay of radon and the ensuing decay of the radon daughters is accompanied by the emission of energy in the form of alpha, beta, and gamma particles. Of these three, alpha radiation is generally regarded as posing the greatest health hazard. Radon emits an alpha particle when it decays into polonium 218. Polonium 218 and polonium 214 both emit alpha particles when they decay. Radon daughters attach themselves to dust particles, which are inhaled and become lodged in the lungs. There the emitted alpha particles can cause cellular damage to lung tissue. Studies on miners indicate that exposure to significant amounts of radon over substantial periods of time can cause lung cancer.

Radon levels are measured in picoCuries per liter of air (pCi/L). A Curie is a measure of radioactivity, and the prefix pico means a multiplication factor of 1 trillion. Another measure is called Working Level (WL), which is a measurement of the potential alpha energy for the decay products of radon. A WL is equal to 1.3×10^5 million electron volts (MEV). Generally, 200 pCi/L = 1 WL.

Radon Detectors

To measure radon levels in a home or building, three types of detectors are generally used—*charcoal canisters, alpha-track detectors,* and *electronic perm (EP) detectors.* Other methods exist, but they are expensive and are typically used only in research.

Charcoal canisters and short-term EP detectors are useful for measurement periods of 3 to 7 days. They are used to secure a quick initial determination of radon levels, called a screening measurement. Radon levels in homes can fluctuate substantially, even within an hour or a day. They also tend to fluctuate seasonally; radon levels are often higher in winter than in summer, sometimes by a factor of two or three. Charcoal canister and short-term EP detector readings, therefore, do not provide an indication of the average annual radon level, which is the critical measurement from the standpoint of health hazards. When additional measurements are needed, alpha detectors or long-term electronic detectors, which provide readings over longer periods of time, are used.

Short-term EP detector measurements that fall within certain broad parameters can be used to determine whether or not additional measurements should be taken as indicated in the following paragraphs.

The EPA publication *A Citizen's Guide to Radon* states that initial screening measurements should be made in the livable area at the lowest elevation of the home—the basement, if the house has one. All windows and doors should be closed for at least 12 hours prior to the start of the test and kept closed as much as possible throughout the testing period.

The Radon Abatement Act of 1988 defines an elevated indoor reading as any reading that exceeds ambient outdoor levels. The EPA's *Citizen's Guide*, which was published before passage of the Radon Abatement Act, used 4 pCi/L (0.02 WL) as a benchmark radon level. This benchmark is reflected in the first of the recommendations regarding followup measurements that appear in the *Citizen's Guide* and are reproduced here:

- If the screening measurement result is less than about 4 pCi/L, followup measurements are probably not required. If the house was substantially closed up prior to and during the test period, there is little chance that the annual average will exceed 4 pCi/L.

- If the screening measurement result is between 4 and 20 pCi/L (0.02 and 0.1 WL), retest with detectors exposed for 1 year, or make measurements of no more than 1 week's duration during each of the four seasons.

- If the screening measurement result is between 20 and 200 pCi/L (0.1 and 1 WL), retest with detectors exposed for no more than 3 months, with windows and doors closed as much as possible.

- If the screening measurement is greater than 200 pCi/L (1 WL), retest with charcoal detectors exposed for no more than 1 week with windows and doors closed as much as possible. The homeowner should consider taking immediate steps for mitigation.

Taking Measurements

Carefully follow directions accompanying charcoal canisters. If one canister is used, it should be placed in the lowest living area of the house—the basement (if the house has one). A second canister is often placed in another area. After a period specified in the instructions, typically 3 days, the canisters are resealed and sent to the distributing firm or laboratory for analysis.

As with charcoal canisters, instructions accompanying alpha-track detectors and EP detectors should be followed carefully. The detectors should be placed in at least two lived-in areas of the house.

A number of commercial firms and laboratories offer radon detection services. The EPA issues a quarterly list of testing organizations (agencies and businesses who make radon measurements) who have voluntarily met EPA standards. Firms are placed on the list after meeting certain EPA proficiency standards. Copies of the list can be obtained from state agencies or regional EPA offices.

You can learn if high levels of radon have been found in your area from the state agencies.

Lead

The two principal sources of lead contamination in the home are lead-based paint and drinking water containing small quantities of dissolved lead.

Studies have shown that ingestion of even minute amounts of lead can have serious effects, including damage to the brain, kidneys, and red blood cells; hypertension in middle-aged men; and injury to the fetuses of pregnant women.

The American Academy of Pediatrics (AAP) regards lead as one of the foremost toxicological dangers to children. Until recently, it was believed that children primarily ingested lead from lead-based paint by chewing chips of old paint or chewing surfaces, such as the protruding edges of window sills. Continued study, however, indicates that dust containing particulate lead in homes where lead-based paint is present is probably a more common source. As with asbestos fibers, these particles can be extremely small, can float in the air as dust, and cannot be fully gathered up by conventional household cleaning methods.

According to the AAP, even low levels of ingestion can cause children to suffer partial loss of hearing, impairment of mental development and IQ, growth retardation, inhibited metabolism of vitamin D, and disturbances in blood formation. Studies have traced intellectual impairment to very low levels of lead in the blood. A report released by the EPA in 1986 said that as many as 250,000 children have suffered measurable IQ losses as the result of drinking lead-contaminated water.

Lead-Based Paint

Lead-based paint was widely used in residential applications until the early 1940s and continued to be used to some extent, particularly for dwelling exteriors until 1976. In 1971, Congress passed the Lead-Based Paint Poisoning Prevention Act. In 1976, the Consumer Product Safety Commission (CPSC) issued a ruling under the Act that limited the lead content of paint used in residential dwellings, toys, and furniture to 0.06 percent. Lead-based paint is still manufactured today for applications not covered by the CPSC ruling, such as paint for metal products, particularly those made of steel. Occasionally, such lead-based paint (for example, surplus from a shipyard) gets into retail stores and the hands of consumers. A study conducted for the EPA in 1986 indicated that there are about 42 million U.S. homes that still contain lead-based paint.

There are several techniques for determining the presence of lead-based paint, including the use of an X-ray fluorescence analyzer. Such tests should be performed by qualified professional personnel. If in doubt regarding the possible presence of lead-based paint, assume that it is present, either as an outer coating or as a coating that has been painted over.

A report issued by a Task Force on Lead Based Paint in Housing, convened in 1987 by the National Institute of Building Sciences, states

> The most serious drawback of present abatement methods [that involve removal of the paint] is the enormous amount of particulate lead, including particles in the respirable range, which is created and can be disseminated throughout the dwelling. . . .Whether micron-sized particles can be removed from old housing surfaces with ordinary vacuum cleaners and scrubbing is extremely doubtful, particularly if splintered flooring is not sealed, covered or replaced.

A number of factors enter into the cost equation for the abatement of lead-based paint. In addition to the present degree of deterioration, the homeowner should examine the structural soundness of the painted surface, the presence or potential presence of water damage that could accelerate deterioration, and, for reasons cited above, the condition of the flooring.

The U.S. Department of Housing and Urban Development (HUD) listed three approaches to abating lead-based paint as being acceptable. They are

- covering the painted surface with wallboard, a fiberglass cloth barrier, or permanently attached wallpaper,

- removing the paint by scraping or heat treatment, and

- replacing the entire surface to which lead-based paint has been applied.

HUD prohibits machine sanding, use of propane torches, washing, or repainting as methods for lead-based paint abatement in housing that it owns and operates.

It should be noted that removing lead-based paint by scraping or heat treatment does not fully solve the problem posed by dust containing particulate lead.

Reporting on a study of abatement measures being tested in Baltimore, Maryland, the National Institute for Building Sciences (NIBS) Task Force stated that

- traditional abatement (open-flame, burning, sanding) resulted in significant increases (typically 10- to 100-fold) in lead levels in household dust, and

- alternative methods used by the city crews (heat guns, thorough cleanup, and repainting) represent modest improvement; however, they did not adequately reduce the hazard associated with dust containing particulate lead.

HUD is currently developing guidelines and recommendations for removing residential lead-based paint. Further discussion of alternatives and techniques appears under "Lead" in the "Rehabilitation" section.

Lead in Drinking Water

The Safe Drinking Water Act of 1986 bans the use of lead pipe and lead solder in drinking water systems. States carry out inspection and enforcement activities. Under the law, the Veterans Administration (VA) and the Federal Housing Administration (FHA) will not provide loans or loan guarantees for new homes or to home rehabilitation projects if lead pipe or lead solder have been used in the construction.

There remain, however, millions of American homes with lead pipes in their plumbing or copper pipes that have been joined with lead solder. Lead pipes are most likely to be present in housing built prior to about 1930. As copper pipes replaced lead pipes, lead soldering of copper pipes was prevalent until passage of the Safe Water Drinking Act of 1986.

When water leaves a treatment plant, it is not polluted with lead. Water is distributed to communities through trunk systems called water mains, which are rarely made of lead. It reaches each home through service connectors, which may be made of lead, and passes through the plumbing to reach individual taps (Fig. 1). Lead contamination may occur when water passes through lead pipes, or through copper pipes that have been joined with lead solder, in the process of distribution to homes or in the home piping.

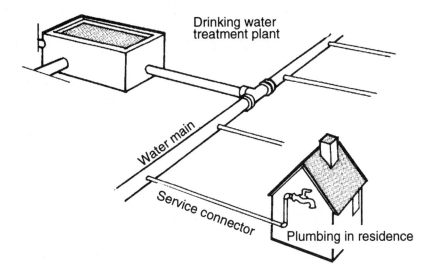

Figure 1—Drinking water supply system.

Lead service connectors can, in some instances, contribute to lead in the water supply. Where this happens, the amount of contamination can sometimes be reduced by raising the alkalinity of the water supply. The homeowner can discuss this with the local water utility and local and state environmental health authorities. However, initial efforts should be directed at the home since most contamination probably occurs in the home plumbing.

Lead contamination of drinking water results from a corrosive reaction between the water and lead pipes or lead solder. In this reaction, small amounts of lead that are invisible and tasteless enter the water.

The amount of corrosion and the resulting level of contamination are functions of several factors, including the chemical characteristics of the water. Water with low pH levels—this is the chemist's way of describing water with a high acidity quotient—will attack lead more readily than water with high pH levels (water with high alkaline content). Soft water generally has low pH levels and thus will attack and absorb lead more readily. However, the process is subject to many variables, and unacceptable levels of lead can be found in water of any hardness.

Particularly high rates of contamination can occur in new copper plumbing systems that have been soldered with lead during the initial years after installation. These plumbing systems were installed either in new homes or as a repair or retrofit in older homes. The amount of contamination decreases as interior deposits form, covering the soldered joints. After about 5 years, contamination from lead-soldered joints can drop to acceptable levels.

In many older homes, particularly those built in the 1930s or earlier, the entire plumbing system may consist of lead piping. Despite the fact that such plumbing has been in place for much more than 5 years, it can continue to pose a contamination threat. Complex reactions that occur with lead solder and result in reduction of contamination over time may not occur in lead piping. In addition, the prevalent practice of using water piping to ground electrical appliances can increase corrosion. Even though a protective scale may build up gradually in lead water pipes, it will not always form, or it may be uneven or may chip in various places.

A laboratory analysis of drinking water is required to determine the presence and amount of lead contamination. Local water utilities, local health departments, and state health departments can often provide information and assistance. In some instances, one or more of these agencies will conduct the test; in others, they refer homeowners to qualified testing laboratories. Some laboratories send a technician to take water samples, or they will instruct the homeowner in the correct method for taking the samples.

Additional information appears in *Lead and Your Drinking Water*, a publication issued by the EPA.

Asbestos

Asbestos is a fibrous mineral substance that is noncombustible, possesses high tensile strength, and has thermal and electrical insulating properties. It has been mixed with various kinds of binding materials to create approximately 3,000 different commercial products, including a number of materials used in home construction. From about 1940 to 1970, it was the material of choice for both insulation and fire protection.

Problems do not occur as long as asbestos and asbestos-containing products remain intact. Problems arise when such products are damaged, broken up, or disturbed in such a way as to cause asbestos fibers to be released. Asbestos fibers are too small to be seen. Their size, as compared with that of other types of small particles, is shown in Figure 2.

The fibers can float in the air for a number of hours and can be stirred up again by ordinary household activity after they settle. They can pass through the filters of ordinary household vacuum cleaners.

There is no current evidence that asbestos fibers can penetrate the skin. However, when asbestos fibers are inhaled and enter the lungs, they become lodged in the lung tissue where they can remain for long periods of time. They can cause serious lung diseases, including lung cancer; mesothelioma, a cancer of the membrane surrounding the lung; and asbestosis, a noncancerous condition in which the lung tissue becomes scarred and loses its ability to function. Symptoms of asbestos-induced respiratory disease usually do not appear for 20 years or more after initial exposure to the fibers.

Persons involved in rehabilitating older homes should know the difference between *friable* and *nonfriable* asbestos-containing materials (ACMs). *Friable ACMs* are products that, when dry, can be crumbled, pulverized, or reduced to powder by hand pressure, causing them to emit

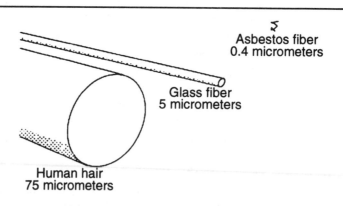

Figure 2—Diameter of asbestos fiber compared with that of other fibers. Reprinted with permission from "Asbestos Handbook for Remodeling: How to Protect Your Business and Your Health" © 1989 Home Builder Press, National Association of Home Builders, 1201 15th Street NW, Washington, DC 20005, (800) 223-2665. p. 4.

fibers into the atmosphere with relative ease. *Great care should be exercised to avoid agitating, crumbling, or otherwise disturbing friable ACMs that are found in older dwellings.*

Nonfriable ACMs contain asbestos fibers that are bound into some type of matrix and cannot escape under ordinary conditions of use for which the products are designed. If they are not broken, torn, sanded, abraded, or otherwise handled in a way that will release fibers, nonfriable materials should not be sources of significant emission. Following is a discussion of some ACMs.

Vinyl Floor Tiles and Vinyl Sheet Flooring

Asbestos can be present in both vinyl floor tiles and vinyl sheet flooring; if it is present, the products fall into the category of nonfriable ACMs. As noted earlier, asbestos fibers can be released if such materials are sanded or damaged, the backing on the sheet flooring is scraped or sanded, the tiles are badly worn, or the products are cut to fit into place.

Patching Compounds and Textured Paints

In 1977, the CPSC banned asbestos-containing patching compounds, and in 1978, manufacturers of textured paints containing asbestos removed these products from the market. They can still be found in older homes. Sanding and scraping of asbestos-containing patching compounds or cutting or sanding of surfaces covered with asbestos-containing textured paint can release asbestos fibers.

Walls and Ceilings

A few homes that were built or remodeled between 1945 and 1978 may contain a crumbly ACM that has been sprayed or troweled onto ceiling or wall surfaces.

Homes constructed between 1930 and 1950 may have asbestos-containing wall and ceiling insulation that is sandwiched behind plaster walls.

Both of these materials are friable ACMs.

Stoves and Furnaces

Asbestos-containing cement sheets, millboard, and paper have been used as thermal insulation to protect the floor and walls around wood-burning stoves. If the label on the cement sheets is present, it may indicate whether the product contains asbestos or not.

Oil, coal, or wood furnaces with asbestos-containing insulation may be found in some older homes.

Insulation for Hot Water Pipes

Asbestos-containing insulation for hot water and steam pipes was manufactured from 1920 to 1972. Such insulating materials include preformed segments made to fit around various diameter pipes and asbestos paper tape. These materials are friable ACMs.

Roof, Shingles, and Siding

Some roofing shingles, siding shingles, and sheet siding have been manufactured with Portland cement that contains asbestos. These materials are nonfriable ACMs.

Asbestos Identification

The presence of asbestos in various products cannot be determined by visual inspection. Experienced home construction personnel can often make a reasonable judgment regarding the presence of asbestos in specific materials found in older homes. However, the safest course is to assume that asbestos is present.

If a major renovation will involve exposing or removing substantial amounts of material behind a wall or other barrier, the materials should be analyzed for asbestos content. The CPSC operates a hotline to provide advice about matters relating to asbestos, including information on sampling and laboratory analysis and assistance in locating a laboratory that can conduct such an analysis. The hotline number is 800–638–CPSC.

Unless they are properly trained and accredited as asbestos abatement professionals, persons involved in the rehabilitation of older homes should leave friable asbestos alone. Even nonfriable asbestos, if it must be removed, requires special handling. Asbestos abatement is discussed under "Asbestos" in the "Rehabilitation" section.

Fire Safety

The danger to life and property from fire in homes is discussed under "Fire Safety" in the Introduction. A number of factors must be assessed in determining the present fire safety of the structure and what might be done during rehabilitation to decrease fire hazards and increase safety.

The interior arrangement of the house may affect ease of egress in case of fire and should be evaluated. Valuable time can be gained for fleeing from a fire if an alarm is sounded early in the buildup of the fire. For example, smoke detectors have been very effective, and their presence and condition should be assessed. Sprinkler systems can decrease the hazard to life and property and may be desirable, particularly in large and expensive homes. The condition of the heating system should be assessed with particular attention paid to coal- and wood-burning stoves and factory-built fireplaces. And finally, special attention should be paid to the electrical system, particularly in very old homes. The system may become overloaded because of inadequate capacity or too few circuits.

House Design and Layout

In redesigning the home's layout, consider the availability of emergency egress for older persons and persons with limited mobility who currently reside in the home or may reside there in the future. Bedrooms on the first floor with easy access to an exit are among the options that can be reviewed.

Smoke Detectors

Battery-powered single-station smoke detectors for residential use were introduced in 1970. They soon displaced heat detectors as the primary device for fire detection in the home. Smoke detectors achieved rapid acceptance in the mid-1970s and have been installed in three of every four homes.

Smoke detectors are widely regarded as the most effective fire safety innovation of recent times, and most analyses estimate that their presence cuts the risk of fire death by half. In some jurisdictions, installation of smoke detectors is required in connection with a permit to carry out a rehabilitation project. Specific types of detectors are required in some local codes, which should be understood before detectors are purchased. Even if they are not required by law, the installation of smoke detectors should be considered a mandatory part of any rehabilitation plan.

Sprinkler Systems

The homeowner may wish to consider installing a fire sprinkler system as part of the rehabilitation program.

A few localities in the United States require that sprinkler systems be installed in new residential construction. As noted in the "Regulations, Codes, and Planned Development" section, some localities also specify that rehabilitation efforts involving half or more of the existing dwelling are covered by regulations governing new construction. If the proposed renovation is extensive, learn whether the local code requires installing a sprinkler system.

Coal- and Wood-Burning Stoves and Factory-Built Fireplaces

With their recent increased popularity, coal- and wood-burning stoves for supplementary heating, and occasionally for primary heating, have become a major contributor to home fire hazards.

An older home that is being assessed for rehabilitation may contain a woodstove that has an unknown installation history. As a first step, the stove should be inspected for cracks, which may permit dangerous quantities of carbon monoxide to escape. Contact the local building inspector or fire department to ascertain if an installation permit was issued, and the proper official should be requested to come to the house to inspect the stove. These steps are not only essential for safety, but documentation may be essential for insurance reimbursement in case of fire.

Guidelines for installing wood-burning stoves and factory-built fireplaces appear in the section on fire safety in the "Hazard Control" section. Review these guidelines when assessing such equipment that is already in place.

Electrical Wiring   The wiring in many older homes is inadequate for modern electrical loads and does not meet modern safety codes. As part of an assessment for rehabilitation, the wiring of an older home should be inspected by a professional electrician. Local regulations may require modernization of the electrical system when a house is rehabilitated.

Indications that wiring should be replaced include the following:

- Fuses that blow or circuit breakers that trip regularly.

- Deteriorated wiring insulation. Sometimes exposed wiring can be found running along the basement ceiling in older homes. Old, nonplastic insulation will often crumble when rubbed between the fingers. In some instances, wires were joined by simply twisting them together.

- Ball-and-tube devices for passing wires through beams and cylindrical porcelain insulators with a nail through the middle for attaching wiring to beams and other surfaces are signs of old wiring. These devices can sometimes be seen in basements or attics.

Fuse boxes are not state-of-the-art electrical equipment for homes. If they are present, replacing them with circuit breakers during rehabilitation is good practice.

Many older homes have insufficient outlets for modern living. This results in "octopus" outlets with a number of extension cords, which is a serious fire hazard. Additional outlets should be included as part of the rehabilitation assessment. It is particularly important to ensure that there are adequate outlets and adequate power in the kitchen, where heat-producing electrical appliances draw substantial amounts of current.

Most homes have some aluminum wiring. Cables bringing current into the home and cables providing 240 V of current to major appliances are likely to be made of aluminum but do not pose abnormal fire hazards.

The fire hazard associated with aluminum wiring is in the "lower branch" of the home electrical system, which provides standard current to lights, switches, outlets, and minor appliances. Aluminum wiring was used in lower-branch circuits from approximately 1965 to 1972, principally in suburban tract homes. There are about 2-1/2 million homes that still contain lower-branch aluminum wiring.

For various reasons related to its physical characteristics, both as a metal and as a conductor, aluminum wiring connections in the lower branch are more likely to heat up or become loose than are copper wire connections, causing greater danger of fire. The CPSC states that the likelihood of an aluminum-wired electrical connection reaching a fire hazard condition is 55 times greater than for a copper-wired connection.

It is good practice to check for aluminum lower-branch wiring when rehabilitating a home that may have been wired or rewired from 1965 to 1972. If it is present, it should be modified to reduce the risk. A professional electrician should determine the presence and extent of the problem and methods to correct it.

Security

Door and window locks throughout the house should be checked and replaced if needed.

The presence and adequacy of exterior lighting should be reviewed for increased nighttime security. Most homes have a light at the front entrance, but this is often the only exterior light. Additional lights can be added to increase coverage of the perimeter, and a timer can be installed that will turn exterior lights on and off when no one is at home. Additional security can be attained by adding deadbolts to entrance doors and window stops to all windows so that the sash cannot be opened widely enough for an intruder to enter.

Thermal Assessment and Moisture Control

This section discusses heat transfer and moisture control as well as different types of insulation, vapor retarders, and signs of excess moisture.

Current heating and cooling costs were not anticipated when older homes were designed and built. As heating and cooling costs increased and homeowners sought more indoor comfort, many homes were retrofitted with various types and amounts of insulation and made more airtight. This often resulted in lower ventilation rates and higher indoor humidity. Vapor retarders were seldom added. This has increased the likelihood of condensation problems that can cause decay and paint peeling.

Since the thermal resistance of a house has become an important economic consideration, many utility companies, contractors, and others perform computerized energy audits to help estimate the savings from energy-saving techniques.

Climatic and Site Considerations

Climatic conditions vary from cold in the northern parts of the United States to nearly tropical along the far southern coastal zones of the United States. As these extremes are approached, assessing factors such as general tightness of the house and the amount and type of insulation becomes important if heating and cooling costs are to be controlled. Climatic conditions also determine where vapor retarders should be placed in order to control condensation.

The orientation of the house in relation to the prevailing winds and sun can have an effect on its thermal efficiency. Trees near and around the house can provide shade or serve as wind breaks.

Heat Transfer

Insulation prevents the conduction of heat from the warm interior of the house to the cold outside in winter months and, if the house is air-conditioned, reduces the warming of the cool inside during hot summer months. Heat may also flow or be transferred from warm areas to colder areas by convection and radiation.

Conduction is the transfer of heat through a material. The denser the material, the higher the rate of heat flow caused by conduction. Good insulating materials are very light with many airspaces.

Convection is a circulatory process in which heated air rises while cooler air moves down. Thus, air near the ceiling of all rooms or on the second floor of a home is warmer than that near the floor or on the first floor.

Heat is also transferred by radiation. Radiant heat is felt when sitting in the sun or by a warm woodstove.

The R-value of materials is a measure of their resistance to conductive heat flow. The higher the R-value, the higher the insulating value. The R-values for common insulating materials are given in Table 3.

Types of Insulation

Mineral fiber, also called rock wool or fiberglass, is the most common type of home insulation. It is composed of fine inorganic fibers made from rock, slag, or glass, with other materials added. Batts and blankets made from mineral fiber are often the easiest and most economical way to add insulation to side walls before closing them in. Such insulation can be used to fill standard-size cavities in ceilings or floor spaces. It can also be cut to fit nonstandard sizes. The batts and blankets are available in thicknesses of 1 to 6 in. and in widths to fit 16- or 24-in. stud spacings.

Batt and blanket type insulation may have a kraft paper or aluminum foil vapor-retarding face with stapling flanges. The other face is sometimes covered with a "breather" sheet. Unfaced batts and blankets are also available; they are held in place by pressure or a friction fit.

Table 3—Thermal resistance (R-values) for selected insulating materials[a]

Material	Density (lb/ft^3)	R per inch	Approximate thickness required (in.)[b]	
			R19	R38
Loose fill				
Mineral fiber (rock, slag, or glass)	0.6–2.0	2.2–2.9	6.5–8.8	13–17
Cellulose	2.3–3.2	3.1–3.7	5.1–6.1	10–12
Vermiculite, exfoliated	4.0–6.0	2.3	8.3	17
	7.0–8.2	2.1	9.0	18
Perlite, expanded	2.0–4.1	3.3–3.7	5.1–5.8	10–12
	4.1–7.4	2.8–3.3	5.8–6.8	12–14
	7.4–11.0	2.4–2.8	6.8–7.9	14–16
Batt/blanket				
Mineral fiber (rock, slag, or glass)	0.3–2.0	2.9–3.4	5.5–6.5	12–13
Rigid board				
Expanded polystyrene, extruded (smooth skin)	1.8–3.5	5.0	3.8	7.6
molded beads	1.0–2.0	3.8–4.4	4.3–5.0	8.6–10
Cellular polyurethane or polyisocyanurate, unfaced	1.5	5.6–6.3	3.0–3.4	6.0–6.8
Cellular polyisocyanurate, gas-permeable facers,	1.5–2.5	5.6–6.3	3.0–3.4	6.0–6.8
gas-impermeable facers, glass-fiber-reinforced core	2.0	7.2	2.6	5.3
Cellular phenolic closed cell	3.0	8.2	2.3	4.6
open cell	1.8–2.2	4.4	4.3	8.6
Mineral fiberboard	16–21	2.7–2.9	6.6–7.0	13–14
Wood fiberboard semirigid	10	3	6	13
structural	10–27	2.1–2.5	7.6–9.0	15–18

[a]R is measure of insulating value or resistance to heat flow.
[b]R19 is thermal resistance often required for walls;
R38 is thermal resistance often required for ceiling.

21

Loose-fill insulation is also available and is packaged in bags. Common insulating materials such as cellulose, mineral fiber, vermiculite, and granulated cork are available in this form. Loose-fill insulation is well adapted for use in ceilings of existing buildings. Precautions must be taken, however, when it is used in side walls since it may settle, leaving some parts of the wall inadequately insulated; this can increase the risk of surface condensation.

Insulation boards made from polystyrene, polyurethane, other plastics, or mineral fiber can act as vapor retarders. They are commonly used as exterior wall sheathing. Some are also suitable as foundation insulation and perimeter insulation for slab-on-grade construction. Insulation boards must be protected from physical damage. Manufacturers' recommendations for installing insulation board on spaced supports in a ceiling, roof, or wall should be followed.

Vapor Retarders

Vapor retarders are designed to keep water vapor out of walls and roofs to minimize or prevent condensation and moisture damage. Visible or concealed condensation in walls and ceilings can cause problems that range from mold growth and paint failure to decay and structural damage.

Figure 3 shows an extreme example of moisture damage caused by winter condensation. The concept of vapor barriers, as vapor retarders were originally called, is approximately 50 years old and was based on the understanding of moisture movement in buildings at that time. A vapor retarder is generally defined as a material with a permeance of 1 perm (a measure of permeability to water vapor) or less, intended to stop moisture movement by water vapor diffusion only. Their use for this purpose was generally accepted in building practice and recommended or required in building codes. Until quite recently, remedial moisture control measures for older homes without continuous vapor retarders focused on water vapor diffusion control by applying paints that retard vapor.

In the 1960s, the importance of air leakage began to be recognized. It became clear that the amount of water vapor carried by air currents could be much larger than the amount delivered

Figure 3—Excessive cold weather condensation on house without vapor retarder on interior walls. Water condensed on the back of siding and then drained through to the outside and froze. Paint blistering and peeling have occurred, and wood decay is likely.

by water vapor diffusion. Now there is little doubt that many of the condensation problems in buildings are associated with air leaks rather than vapor diffusion. If vapor retarders had been effective in preventing condensation, it was primarily because continuous vapor retarders usually also limit air flow.

Although vapor retarders are no longer widely considered as the best way to prevent moisture condensation in walls and roofs, no new consensus on the most effective methods has emerged. A discussion of other ways to control moisture and condensation follows in the "Rehabilitation" section. For lack of widely accepted guidelines, this discussion of vapor retarders is based on past and current practice.

Traditionally, recommendations for vapor retarders have been based on climate. Three different climate zones are recognized in the United States, based on winter design temperatures (Fig. 4). Zone I includes areas with design temperatures below -20°F, zone II between 0°F and -20°F, and zone III above 0°F with exception of areas with extreme warm humid summer conditions. Within each zone, similar types of condensation problems are expected and similar corrective measures are recommended.

Exact guidelines describing which buildings require a vapor retarder are not currently available. Vapor retarders are traditionally recommended in all exterior walls in zones I and II and in zone III when the wall is insulated beyond R4. Ceiling vapor retarders are traditionally recommended for zone I. Ceiling vapor retarders are recommended in zone II as well. These recommendations conform with current interim guidelines from the American Society of Heating, Refrigerating, and Air-Conditioning Engineers (ASHRAE).

The toughness of a vapor-retarder material is as important as its ability to stop water vapor, since a torn vapor retarder has no protective value. Polyethylene is the material most commonly used as a vapor retarder. A thickness of at least 4 mils, preferably thicker, is recommended. Other materials include aluminum and several reinforced barrier materials specially marketed as vapor retarders. Kraft paper is much less effective as a vapor retarder because it has a higher permeability and tears quite easily.

The vapor retarder must be continuous, especially if it also functions as the main air barrier in the wall or ceiling. Penetrations, such as electrical outlet boxes, must be carefully sealed. Vapor-retarder flanges should be stapled to the front face of the stud, not the side of the stud. Flanges should overlap.

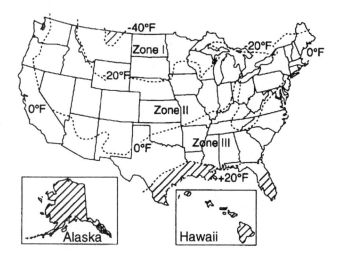

Figure 4—Condensation zones in United States.

Adding vapor retarders in existing buildings without removing the interior finish material is difficult. Vapor-retarding paints can be used effectively, but air leaks must be eliminated. In older buildings, sealing all possible air leaks is often difficult. In such cases, keeping indoor humidity at a safe level is especially critical. This is even more critical if a new flueless energy-efficient or electrical furnace has been installed and the building has been made more airtight. Methods to control indoor humidity are discussed in the "Rehabilitation" section.

Assessment of Insulation

Both the type of insulation and the application method can affect the overall effectiveness. In the assessment, be certain to note (1) the type of insulation present and the R-value (if listed), (2) the thickness of the insulation, (3) the location of the insulation, and (4) the condition of the insulation with regard to closing all voids (quality installation) and any physical damage that has occurred.

Ask the current occupant or owner about insulation. The occupant should also have a good idea of current heating and cooling costs. Ask to see utility bills, which should help to determine the effectiveness of the current insulation and could be of value in determining the return on insulation or other weatherization measures being considered.

Check the attic and ceiling areas for insulation. In pitched-roof houses, these areas are usually accessible, making both the inspection and application of insulation easy. These areas may also be the most effective areas to insulate in order to control energy loss. Make certain the insulation has been installed between and evenly distributed over the tops of the ceiling joists and that it is not blocking air vents, particularly those in the soffit areas. Check for damp wood, wood discolored by water stains, and decay in attic sheathing, rafters, and other wood members (Fig. 5). Water stains on ceilings below the attic area may indicate moisture problems from condensation.

Figure 5—Darkened areas on roof boards and rafters in attic area show where there has been condensation. This could probably be prevented by installing a vapor retarder in the ceiling and by good attic ventilation.

In houses with a basement or crawl space, check for insulation in the foundation walls and in the sill areas. Materials used for foundation insulation must be able to withstand moisture. The most commonly used materials are rigid sheets of polystyrene, polyurethane, or asphalt-impregnated rigid insulations. Often sheet insulation is placed around the perimeter (cold side) of the foundation.

Check for the presence or signs of condensation (Fig. 6) on crawl space walls, sills, and joists. Generally, vapor retarders should be in place over the soil, where they are very effective in limiting evaporation. In addition, local building codes usually require that the crawl space be ventilated.

Once they are closed in, exterior walls are the most difficult and expensive to insulate. Insulation in the wall cavities of older homes probably was blown or foamed into place. In these cases, it is important to ensure that all wall voids are completely filled and that unreasonable settling or shrinkage of the insulation has not occurred. It is extremely difficult to determine if this has happened without literally taking the walls apart. However, a professional inspection with infrared thermography can reveal settling, shrinkage, and other voids in wall and ceiling insulation.

Blown-in or foamed insulation is frequently applied through holes bored in the exterior side walls. The plugs installed in the siding after the insulation is in place frequently are visible. Another application method involves removal of the siding at the top of the wall cavities and above windows, boring the sheathing, and blowing the insulation into place. The siding is then replaced, and there are no holes to indicate that the wall has been insulated.

Figure 6—Surface condensation on floor joists in crawl space. Vapor-retarder ground cover and proper ventilation can prevent this condensation by restricting water vapor movement from the soil, thus preventing high humidity in crawl space and subsequent condensation.

Side walls may be covered with rigid board insulation that is then covered by siding. The presence of insulating materials under the siding is generally apparent from a visual inspection at the bottom edge of the siding.

Airtight House and Foundation

To meet fire code requirements, the interior side of many insulating materials must be covered with 1/2-in. gypsum board or its equivalent. If the insulating material is not covered, check with local building officials to determine what must be done to make the installation safe.

An important factor in thermal assessment and moisture control is whether or not the house is airtight. Drafts from air leaks around windows, doors, and any other openings are often present in older homes and should be corrected. Correction of these problems is often expensive, so make a thorough investigation of the condition of existing windows and doors. For a discussion of window and door assessment, see "Windows and Doors" in the "Condition Assessment" section.

The foundation should be in good repair. Make certain the sills fit tightly on the foundation and that they have not deteriorated or been damaged by insects or fungi. The mortar between joints in block foundations should be tight, not cracked, and not eroded away. Foundations are discussed further under "Foundations" in the "Condition Assessment" section.

Symptoms of Excess Moisture

Although the majority of wood-frame buildings do not have serious moisture problems, such problems are a source of numerous complaints and owner dissatisfaction. Excess moisture occasionally leads to decay and subsequent structural damage.

Most moisture damage is caused by leaks in the roof or water pipes, improper drainage, or wind-driven rain. Serious damage from condensation is an exception. When condensation does occur, it is usually caused by a combination of factors. Condensation during winter is usually due to high indoor humidity and poor construction details.

Many symptoms of excess moisture are readily apparent. Therefore, focusing on less-obvious symptoms is good practice. Some moisture symptoms may be seasonal and may not be evident at the time of inspection. For example, the soil in a crawl space may appear dry in mid-winter; however, during other periods it may be wet from summer rains or poor drainage. Musty odors and the sensation of dampness may signal mold, mildew, or decay. Growths of mold or mildew often produce discoloration with colors such as white, orange, green, brown, and black. Mold and mildew indicate that conditions favoring decay are present. This is discussed in more detail under "Decay" in the "Condition Assessment" section.

Discolorations, stains, and texture changes indicate that the wood has been wet at some time. Although the area may not be wet at the time of inspection, such symptoms suggest that the area should be carefully inspected for damage, especially from decay organisms.

Condensation on windows and other cold surfaces, such as water pipes, can indicate excessively high humidity. Excessive humidity may result from two causes—inadequate ventilation of the area or production of an inordinate amount of water vapor. High humidity is most likely to occur in kitchens or bathrooms; exhaust fans in these areas should be checked to be sure that they are operating properly and are of sufficient capacity. In the basement, clothes dryers should be checked to be sure that they are vented to the outside and that the vent is not blocked. If room or furnace humidifiers are used during the winter, make sure that they are set at the proper humidity level. There may be excessive moisture when the dwelling has an unusually large number of occupants, necessitating more than usual cooking and laundering activities.

Figure 7—Ice dams in cold climates can cause severe moisture problems in roofs, ceilings, and walls. Peeling and staining of paint may be summertime evidence of problems.

Frost or ice on any surface can indicate trouble. Roof ice dams frequently occur in roof valleys (where two roof lines meet) and along the eaves (Fig. 7), when there is insufficient ceiling or attic insulation, there is poor attic ventilation, or the attic is used as living space and heated. Snow or ice melts as the sun strikes the roof or as the roof surface is warmed from within. This moisture runs down the roof and freezes at the roof edge where the roof surface is colder, forming dams that impede drainage. As the freezing and thawing cycle continues, the condition worsens. Ice can build up under shingles, or moisture can back up at the edge of the roof. Water then leaks into the attic or runs into the wall cavities.

Damage to exterior paint may be a result of moisture problems from either water on the exterior surfaces or water vapor migrating out from the interior. Paint that has peeled or blistered, exposing the surface of the substrate, is a signal of paint damage caused by moisture. This condition is discussed in detail under "Exterior Finishes" in the "Condition Assessment" section and under "Application" in the "Rehabilitation" section.

After moisture moves through concrete or masonry (such as a brick veneer wall), a white powdery substance known as efflorescence may form or moisture lines may appear.

The Site

In this and the 14 ensuing sections, you will be taken on an inspection tour of a house and its grounds to assess their condition for the purpose of rehabilitation.

Grading

Grading the soil away from the foundation allows water to drain away from the house (Fig. 8). This minimizes moisture penetration through foundation walls or under slab foundations. Frequently there is improper grading behind shrubbery or in flowerbeds adjacent to the house and at the base of downspouts. Improper grading may also be caused by missing or deteriorated roof gutters, which allow water running from the roof to erode the soil.

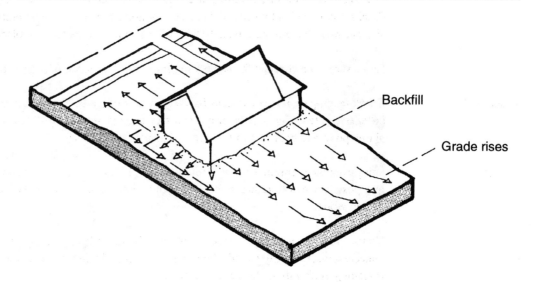

Figure 8—Proper soil grading to drain water away from foundation.

Walk around the perimeter of the house and inspect the slope of the ground where it meets the foundation. Check for evidence of water accumulation or puddles. If a crawl space is present, be sure to check it for evidence of moisture accumulation.

If the house has an outside patio, ensure that it slopes away from the house and that it has no depressions where water can collect.

Landscaping

Common landscaping defects are the absence of trees and shrubbery and poorly established lawns. If you want to improve the external appearance, decide on its priority in relation to other needs, and include such changes in the rehabilitation budget.

In addition to their aesthetic value, trees and shrubs near the house contribute to the microclimate of the home and its immediate external surroundings. They mitigate extremes of summer temperatures and, in turn, contribute to the energy efficiency of the home.

If the work that is to be done on the house involves the movement of trucks or heavy equipment over the grounds, drivers should be instructed to avoid the immediate vicinity of trees, where roots are a short distance beneath the soil, and the homeowner should maintain vigilance to be certain that the instructions are carried out. The damage that can be done to the roots, invisible at the time, can cause the tree to die slowly over a period of several years.

Trees in close proximity to the house may eventually damage the foundation. Tree roots can split rocks, so they can harm foundations as well. If you are in doubt about the potential problems that may be posed by a tree near the house, consult a landscaping or tree specialist.

Walkways, Driveways, and Patios

Walkways and patios are sometimes made from concrete blocks of various forms set directly into the ground. They should be inspected for cracks, chips, or other damage, and for heaving or settlement, which could be hazardous.

Inspect isolation joints between concrete slabs. Cracked or missing caulking should be replaced. A distance greater than 1/2 in. between slabs, or vertical displacement in excess of 1/4 in., is unacceptable, and the homeowner should consider replacing the slab.

Closed hairline cracks are normal and, aside from appearance, should not cause concern. Cracks accompanied by vertical displacement greater than 1/4 in. and cracks without vertical displacement that are more than 3/16 in. wide should be repaired or replaced.

Patios should be checked for uneven settling and for evidence of separation from the house.

Fencing

Check fencing or wooden screens for deterioration of the paint or stain; loose boards and fasteners; rotted rails, boards, or posts; insect damage; settling or frost-heaving of posts; out-of-plumb posts and boards; and missing or damaged gate hardware.

Pay particular attention to areas where wood is in contact with the ground. If posts, particularly gate posts, are deteriorating in the ground, it may be necessary to replace them with pressure-treated wood.

Vertical alignment can be checked visually. A builder's level or plumb bob is generally unnecessary. A slightly out-of-plumb fence—about 1 in. in 6 ft—is not a problem as long as it is leaning uniformly in the same direction.

Interior Layout

When an older house is being rehabilitated, the owner will probably wish to go beyond structural repair and renovation. Rehabilitation provides the opportunity to improve the layout, add more space and modern conveniences, and improve appearance. The layout of an older home may not easily accommodate every improvement in arrangement and convenience that the homeowner may desire, but it will probably accommodate some of them.

The owner should survey the house, note all desirable changes, and list them in order of priority. Decisions can then be made that reflect both individual wishes and the budgetary considerations cited in the Introduction.

For additional information about assessing future rehabilitation projects, see "Interior Structural Changes" in the "Rehabilitation" section.

Preservation and Restoration

As noted in the Introduction, every house is historic in that it reflects the tastes and the cultural world of its time. When substantial changes in the interior of an older house are considered, questions about preservation and restoration arise that can affect the home's value.

Potential buyers of old homes are usually looking for something other than a standard inventory of measurements and conveniences. That additional something could be called charm. Where charm exists in old houses, caution should be exercised in removing it and simple steps can be considered to enhance it.

Many new house designs borrow features from older styles. These include the dignity of the two-story Colonial, the quaintness of the Victorian house, the charm of the old English cottage, the coziness of the early 20th century bungalow, the solid comfort of the Midwestern farmhouse, or the rustic informality of the old-style ranch house. If an old house possesses any of these or other desirable characteristics, it may be wise to avoid drastic changes of appearance; at the least, certain features can be retained in the context of change and renovation.

Interior features of older houses can include high ceilings, woodwork, molding, plaster walls, and period fireplaces. Such features have sometimes been modified or entirely stripped away in the process of rehabilitation, resulting in changes in basic appearance and proportions that could reduce, rather than enhance, both the interior beauty and the marketability of the house. As an alternative, layouts and rehabilitation plans can be devised that preserve and incorporate such features. Literature and consultants on historic preservation may need to be referred to if the house is located in a historic area.

Accessibility for Special Groups

The layout and interior design of most older homes, and of many homes built more recently, do not reflect the accessibility needs of special groups such as the elderly, the handicapped, and the disabled. In recent years, there has been increasing awareness that rooms and facilities laid out without considering the needs of these groups can pose immense problems of accessibility for them.

Several publications on the subject of accessible housing have been produced by Federal and State agencies, nonprofit groups, and commercial publishers. A publication issued by the Department of Housing and Urban Development (HUD), entitled *Housing Special Populations: A Resource Guide*, lists and describes many of the available printed resources in the field.

Entryway

An entryway or foyer with a light and an adjoining coat closet can be constructed where one did not exist. The entry should be located adjacent to the living room and should provide access to other areas without the necessity of passing directly through the living room, or at least not more than a corner or end of the living room.

Living Rooms, Family Rooms, and Dining Areas

The living room should include one wall that is at least 10 to 12 ft long to accommodate a sofa and end tables. Bathrooms should not be in direct sight.

The primary dining area can be positioned at the end of the living room or offset from the living room, forming an L. Some homes have a room that combines the functions of a dining room and family room, with a dining area adjacent to the kitchen. In either case, the dining areas can be in sight of the food preparation areas of the kitchen. The family room might have an exterior door that opens to a patio or yard.

If the primary eating area is a separate dining room, it should be of adequate size to accommodate comfortably a dining room table and chairs plus a china closet or buffet, or both.

Bedrooms

Houses will typically have a master bedroom plus one or more additional bedrooms. These sleeping areas should be separated from other areas of the house and should have convenient access to bathrooms without passing through or being in sight of other areas of the house. When there are three or more bedrooms, the master bedroom is usually served by its own full bathroom. The master bedroom should be of adequate size to accommodate at least a queen-size bed and other furnishings, with adequate space in hallways and passageways to permit access for such furniture.

Kitchens and Bathrooms

New appliances, concepts of convenience, and living styles have revolutionized the kitchen, even within recent years. In addition to containing older appliances, or failing to contain certain modern appliances such as dishwashers and not providing space for them, older kitchens are likely to be the wrong size and shape for efficient modern use. Some suggested kitchen arrangements are shown under "Remodeling the Kitchen" in the "Rehabilitation" section.

In studying interior layout, consider the high priority that kitchen modernization is receiving. If the prospective owner will be modernizing and changing the size and shape of the kitchen, the changes must be taken into account in laying out the total floor plan.

Older houses are often deficient in bathroom facilities; many have only one bathroom. Additional space for these facilities must be provided in the layout; special consideration must be given to the location of existing plumbing and water lines. Information on adding a bathroom and some suggested layouts appears under "Heating, Air-Conditioning, Plumbing, and Electrical Systems" in the "Rehabilitation" section.

Expansion of Available Space	Regardless of the size of the house, there always seems to be a need for more space. Finishing an attic or basement offers opportunities for internal expansion. These options are discussed under "Attic Space and Shed Dormer" in the "Rehabilitation" section.
Garage	Another place for expansion is the garage, which can be built as an integral part of the house. If the garage is well built, the only work is in finishing, which is much less costly than adding onto the house. The main consideration is whether the additional finished space is needed more than the garage. The garage is often adjacent to the kitchen, which is an ideal location for a large family room. It could also be used for additional bedrooms and possibly another bathroom.
	Walls and ceiling of the garage can be finished in any conventional material. The floor will probably require a vapor retarder, insulation, and a new subfloor. Using the existing garage door opening to install large windows may be convenient; otherwise it can be completely closed and windows added at other points. Details of converting a garage to other uses are covered in the "Rehabilitation" section under "Conversion of Garage to Living Space."
Additions	If space requirements are still not met after considering all the possibilities of expanding into the attic, basement, or garage, the only alternative is to construct an addition. The house on a small lot may have minimum setback limitations that will present problems. Local zoning and lot restrictions should be checked. The distance from the front of the house to the street is usually the same for all houses on a particular street, so expanding to the front may not be permitted. Often a house also has the minimum setback on the sides, preventing expansion on either side. Thus, the only alternatives are to add on behind the house or to add a second story. A house in the country usually can be expanded in any direction without restriction.
	Use the addition for the space most critically needed. If the need is for more bedrooms but a much larger living room is also desirable, maybe the present living room can be used as a bedroom and a large living room can be added. If the main requirement is a large modern kitchen, adding a new kitchen and using the old kitchen as a utility room, bathroom, or some other type of living or workspace may be feasible. The important thing is that the addition be similar in style to the rest of the house. Roof lines, siding, and windows should all match the original structure as closely as possible to give the house continuity, rather than giving the appearance of something that has just been stuck on.
	One of the most difficult problems is connecting the addition to the original house. Sometimes the satellite concept is used, by which the addition is built as a separate building and connected to the original house by a narrow section that could serve as an entry or mudroom and includes closets or a bathroom (Fig. 9).
	One disadvantage of this approach is the resulting large exterior wall area with proportionate heat loss and maintenance cost for exterior surfaces.
Layout Change of More Recent Homes	This handbook is directed principally to rehabilitation of older housing. However, houses built after 1945 sometimes require rehabilitation, and the interior layout of these homes may also require modification when rehabilitation is being planned.
	Figure 10 shows suggested changes in the layout of several types of post-1945 houses. These plans may also provide ideas for modifying the layout of older homes.
Foundation	The foundation is generally a concrete or masonry structure that supports the house, and its soundness is essential for successful rehabilitation. Foundations should be checked for general deterioration that may allow moisture or water to enter the basement (for information on

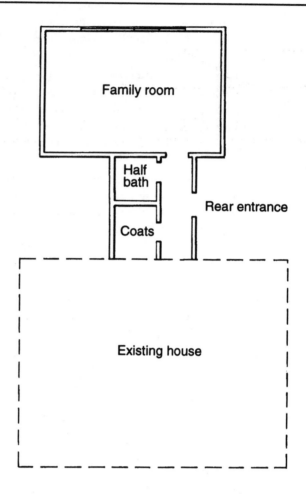

Figure 9—Addition using satellite concept.

basement moisture, see "Basement" in the "Condition Assessment" section. Even more basically, however, foundations must be checked for uneven settlement (Fig. 11), which can distort the house frame and even pull it apart. A single localized failure or minor settling can be corrected by releveling beams or floor joists and is not sufficient reason to reject the house. Numerous failures and general uneven settlement, however, may render the house unsuitable for rehabilitation.

If an assessment that is conducted as described in this section indicates the possibility of significant problems, an evaluation should be secured from a professional building analyst or professional engineer.

A typical residential foundation wall consists of two major elements—the vertical wall of masonry or concrete and the footing on which the wall rests. In older houses, the foundation walls may be made of stone or brick. The footing is made of poured concrete and is usually about 8 in. deep and 16 to 24 in. wide. The concrete slab that forms the basement floor carries no weight and is not a structural part of the foundation.

Normal Settlement

Settling of the foundation usually involves two elements. First, the footings of the house may be unevenly loaded. Second, the footings may rest on soil that has a bearing capacity that is not uniform along the lengths of the footings.

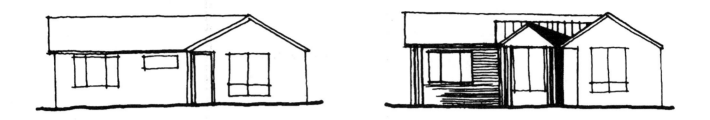

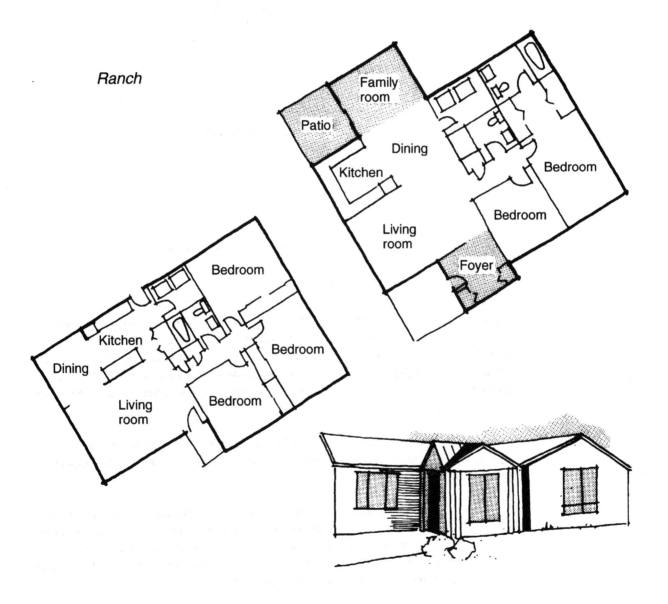

Ranch

Figure 10a—Options for revising floor plans of ranch-style houses with before (left) and after (right) floor plans.

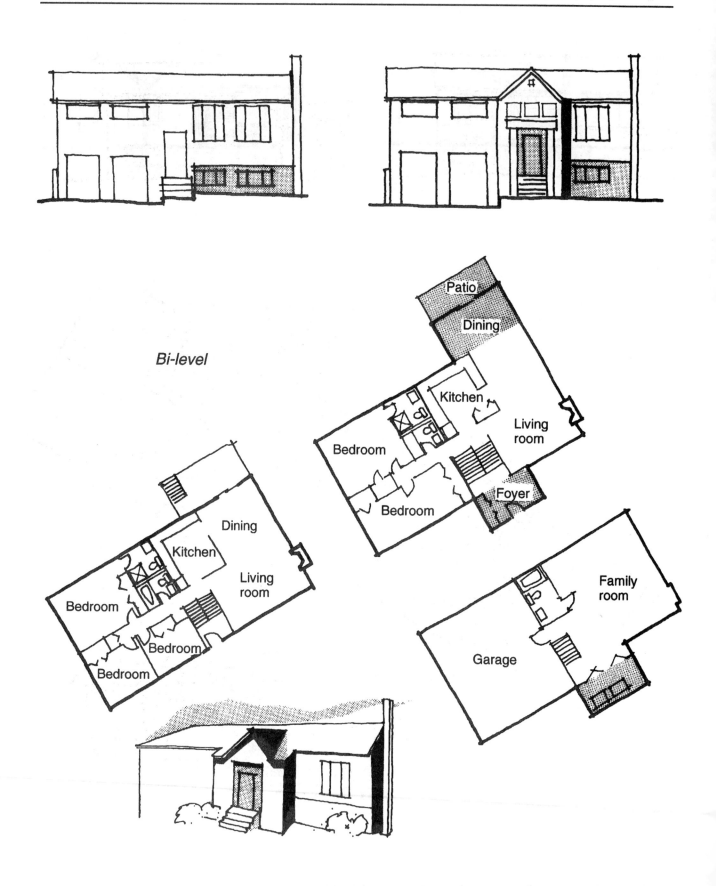

Bi-level

Figure 10b—Options for revising floor plans of bi-level houses with before (left) and after (right) floor plans.

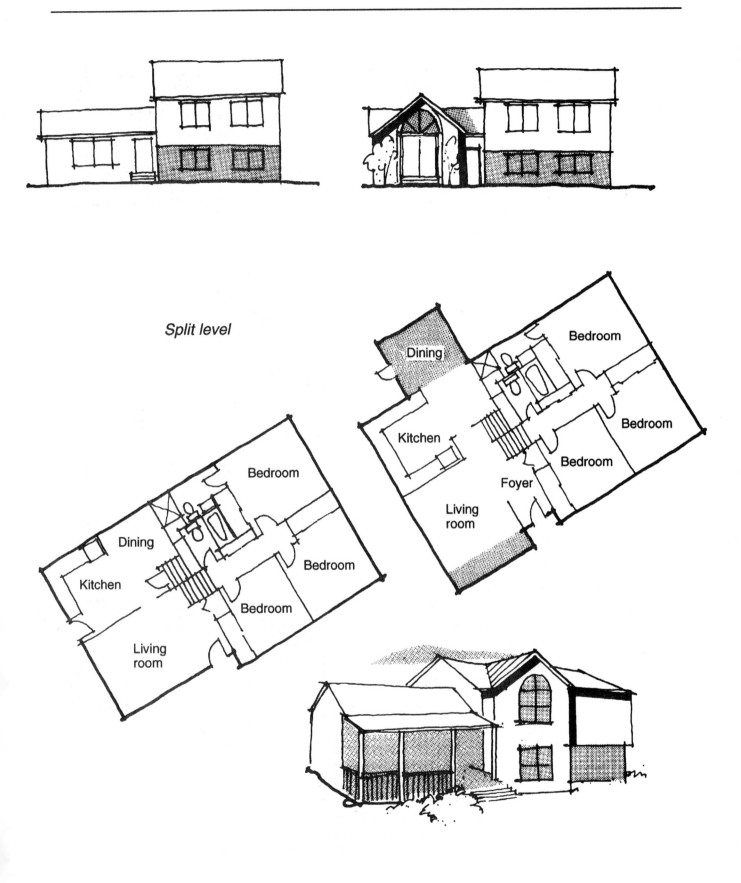

Figure 10c—Options for revising floor plans of split-level houses with before (left) and after (right) floor plans.

Figure 11—Uneven foundation settlement.

When a house is first built, the foundation adjusts to these forces during a period of about 2 years. This adjustment can produce, on a one-time basis, hairline cracks in the foundation walls. With a masonry block wall, such cracks typically extend diagonally from a corner along the "steps" of the blocks, widening as it rises from the floor. Cracks of this type, which were produced during the early period of settling, can usually be safely ignored. Wider cracks and active cracks—those that are continuing to spread and widen—need more thorough investigation.

Soil Moisture and Foundation Settlement

An important factor in foundation distress is moisture. Soil moisture or lack of it, or cyclical wetness and dryness in the soil can cause changes in the bearing capacity of the soil. In addition, wet soil or soil expansion caused by frost can increase pressure against the walls of the foundation and cause them to bulge.

Checking Foundations

When checking the condition of the foundation, inspect for cracks in the vertical wall. As indicated earlier, stable hairline cracks created during initial settling are generally not important. Crack openings in excess of 1/8 in., or cracks with differential settling of the wall sections on either side of the crack, may indicate a serious foundation failure. Potentially serious problems may also be indicated by cracks that show recent or continuous movement. Indications of such movement include

- previously repaired cracks that have reopened,

- cracks on painted surface with no paint in the crack, and

- cracks with newer, sharp edges.

The foundation walls should also be checked for caving or bulging. Attach or hold a string at each end of one side of the foundation wall and pull it taut. Caving or bulging will show as deviations from the line of the string. Caving or bulging in excess of 1 in. along the length of the wall may indicate serious failure.

Each side of the foundation should be checked for corner or center uplift. Attach or hold a string to the top of each foundation wall at the corners. Settling at the corners or along the length of the wall will show as deviations from the line of the string. Displacement in excess of 1/2 in. may indicate serious failure.

Exterior Walls

Exterior walls are important in that they are a major structural element of the house and, in addition, help to determine its architectural style. Important considerations for rehabilitation include the type of siding, its condition, and the finishing method, as well as the placement and condition of windows and doors and the trim around them.

Siding

Exterior siding protects the interior structural members from the weather and also provides some thermal resistance to the walls. The type of siding and the manner in which it is finished determines to a large extent the style and aesthetic appeal of the house. Solid lumber, shingles, brick, and stucco were the most commonly used materials for exterior siding on most older homes. More recently, plywood, hardboard, vinyl, and aluminum siding have become popular.

Wood

If there are problems with exterior wood siding and trim, they probably stem from excessive moisture, which can enter from either inside or outside. Lack of a roof overhang will allow rain to run down the face of the wall. If there is no vapor retarder, moisture can also enter the sheathing or the siding from the inside of the house. Problems such as warping and nail pulling may also be expected if the siding is inadequately dried before it is installed. Rust stains from nailheads may appear unless the nails are properly galvanized or made of stainless steel or aluminum.

The species of wood and the way lumber is sawn from a log will, in part, determine the amount of warping, decay, and finish deterioration of wood siding. The heartwood of redwood, cedar, cypress, and white pine was commonly used for siding because these woods are uniform in texture, durable, lightweight, and tend not to warp or change dimensions excessively. Southern Pine and Douglas Fir have also been used for siding. These woods are heavier and tend to warp and to change dimensions more. Southern Pine manufactured in the 1930s or later tends to be predominantly sapwood and is not naturally durable. The heartwood of Douglas Fir has some natural decay resistance, but not as much as redwood and cedar. Southern Pine and Douglas Fir have a characteristically distinct latewood band (Fig. 12) and are not uniform in texture. Latewood (also called summerwood) is denser, harder, darker in color, and composed of cells with thicker walls and smaller cavities than the earlywood (also called springwood). As a result, paint tends to fail on the latewood. Depending on how a board is cut from a log (Fig. 13), it can be either edge-grained or flat-grained.

Flat-grained lumber tends to expose more of the dense latewood in species such as Southern Pine and Douglas Fir, resulting in a shorter life expectancy for painted surfaces.

To check for loose siding, look for space between horizontal boards by standing very close and sighting along the wall. If the boards are not badly warped, renailing may solve the problem. New siding may be required if general gapping or looseness is evident, or if the finish cannot be restored with reasonable effort.

Check the siding for decay where two boards are butted end to end, at corners, and at window and doorframes (see "Recognizing Damage" in the "Condition Assessment" section). Where new siding of any type has been applied over the original wood siding, check for signs of decay or insect damage in the original siding where it may be visible, at splits or gaps and on window and doorframes.

Figure 12—Paint fails first over bands of latewood in softwoods: (a) two boards that were painted with white paint and exposed to weather for 5 years; (b) the same boards after remaining paint was taken off with paint remover, showing dark bands of latewood, indicated by paint-bare areas in (a).

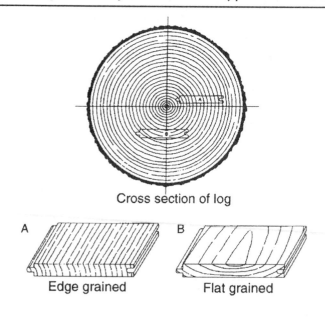

Cross section of log

A
Edge grained

B
Flat grained

Figure 13—Effect of sawing method on ring orientation in lumber.

Shingle	Wood shingle siding in good condition appears as a perfect mosaic. Worn shingles have an overall ragged appearance, and close examination may show individual shingles to be broken, warped, and upturned. Badly weathered or worn shingles should be replaced.

Stucco and Masonry

The exterior of some older homes is covered with siding shingles or sheets made of Portland cement that contains asbestos. As stated in the earlier section on asbestos, these materials pose no problems as long as they are left intact. Problems are associated with removal, especially the possibility that the process may damage the materials and cause tiny particles of asbestos to float in the air and be inhaled. Discussion of techniques for removal of exterior asbestos-containing shingles appears under "Asbestos" in the "Rehabilitation" section.

The exterior of some older homes is finished with stucco. Where the stucco is noticeably stained, or where it is cracked all the way through in a number of places, new reinforcement and stucco will be required. When stucco is applied to platform-framed two-story houses, shrunken joists and sills may cause unsightly bulges or breaks in the stucco unless the moisture content of the framing members was at the proper level when installed or they had reached equilibrium moisture content at the time the stucco was installed.

If brick or stone veneer was used as siding, the mortar may become loose and crumble, or uneven settlement may cause cracks. In either case, new mortar should be applied after any settlement has been corrected. Some older houses were trimmed with soft bricks and porous stone. Painted, stained, or dirty brick and stone may be cleaned by sandblasting, but care must be taken in sandblasting soft stone and brick so that not too much material is removed. In such cases, the brick and stone may need to be treated with transparent waterproofing to maintain the desired appearance.

Other Siding Types

Other types of siding may be found on older homes. For example, panel siding may be in the form of plywood, hardboard, or particleboard, as well as of aluminum or vinyl. In general, the wood-based materials should be examined for decay and other damage as described earlier in this section, and hardboard siding should also be examined for buckling. Nonwood materials should also be examined for mechanical damage, such as dents or tears, and for integrity and color of the finish.

Windows

Windows often present one of the more difficult problems in old wood-frame houses. If they are loose-fitting and not weather-stripped, they will be a major source of drafts and heat loss.

Check the window's tightness and fit, and examine the sash and sill for decay. Examine the condition of the glazing putty and the caulking around the frame. If the wood parts have decayed, they must be replaced in whole or in part. Decay usually begins in the joints of the window. Check the operation of the window. Operate all movable window sashes. One or both sashes of double-hung windows in older houses may be painted shut; they can be opened with a sharp, pointed knife. With both sashes freed, determine if the window latch will operate properly. Check glazing and putty. Casement windows should be checked for warp at top and bottom.

In cold climates, windows should be double-glazed or have storm windows to prevent condensation and reduce heat loss. If the windows are not a standard size, storm windows will probably be expensive because they will have to be custom made.

Check the window dimensions when planning to replace windows. If the window is not a standard size or if a different size is desired, the opening will have to be reframed or a new sash made, both of which are expensive.

Doms

Exterior doors should fit well without sticking. They should be weather-stripped to prevent air infiltration, which can be done inexpensively. Difficulty in latching a door is usually attributable to warping. Adjustment of the latch-keeper will solve the problem in some instances, but badly warped doors should be replaced.

Storm doors are necessary in cold climates for comfort and for energy conservation. They also retard moisture condensation on or in the house door and protect it from severe weather; they should be included in planning.

If the doorframe is out of square because of foundation settlement or other racking of the house frame, the opening will probably have to be reframed.

The lower parts of exterior doors and storm doors are particularly susceptible to decay and should be carefully checked. Also observe the condition of the threshold, which may be worn, weathered, or decayed and may require replacement.

Exterior Finishes

Properly applied and maintained exterior finishes can extend the life of a siding material as well as enhance the overall appearance of the house and help define its style. However, finishes have limited lifetimes, and a buildup of finish (many layers) may hasten failure. Finish failure is usually the result of construction, maintenance, moisture, or application problems, which must be corrected if the new finish is to last.

Chalking

Many older homes were sided with wood and finished with paint. Barring problems such as those just mentioned, paint normally wears by chalking (Fig. 14). Chalking is the slow release of individual particles from the painted surface by weathering or deterioration. The rate of paint chalking depends on formulation and exposure of the paint.

For the application shown in Figure 14, a special nonchalking paint should have been used to prevent the discoloration.

Figure 14—Paints normally fail by chalking, which is the gradual weathering-away of individual paint particles.

*Cross-Grain
Cracking*

Excess paint buildup often causes cross-grain cracking (Fig. 15). The only solution to this problem is to completely remove the old paint, which can be difficult and expensive. If the siding is also in poor condition from warping, checking, decay, or other damage, it may be better to simply replace it or cover it with another siding material.

Peeling

Peeling is another common paint problem. It often results from excess moisture in the wood. Short or nonexistent roof overhangs and improper maintenance of gutters may cause rain to penetrate the siding. The absence of a vapor retarder on the inside of exterior walls and ice dams on roofs can both cause excess moisture in the wall cavities. Excessive peeling (Fig. 16)

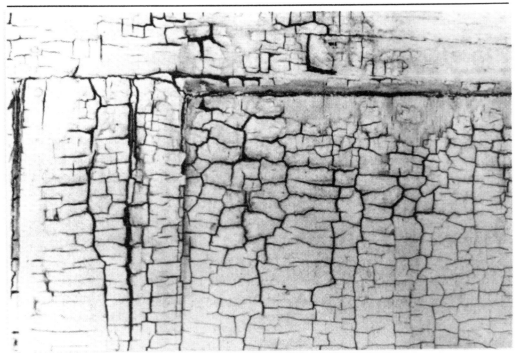

Figure 15—Cross-grain cracking results from excessive buildup of paint.

Figure 16—Paint peeling from wood can result when excessive moisture moves through the house wall. Some cross-grain cracking is also evident on this older house.

may require that the paint be completely removed. The source of moisture must be eliminated or the paint will peel again.

Paint peeling can also result when incompatible coatings are applied successively, the old surface is not properly prepared, the new paint is not properly applied, or an inferior paint is used. The peeling paint must be removed and the surface properly prepared before new paint is applied.

Discoloration

Discoloration of painted surfaces, common on older homes, can have several causes. One cause is mildew (Fig. 17), a common problem in warm, humid climates. If the mildew is not killed before new paint is applied, it will grow through the new coat and discolor it. If mildew is a problem, paint containing a mildewcide may be used.

A run-down or streaked discoloration (Fig. 18) usually happens in cold climates when water condenses on the back side of redwood or cedar siding, dissolves the extractives in these darker woods, and carries them to the exterior surface of the board below. If the moisture problem is corrected, the discoloration will eventually weather away or the surface can be scrubbed clean.

Rust stains can usually be traced to the presence of iron; nails are the most common source. All nails used for exterior purposes should be high-quality galvanized stainless steel or aluminum nails. If not, rust stains will occur.

Safety Considerations

Lead-based paint was commonly used on the exterior of dwellings until 1976. It is safe to assume that any exterior paint applied before this date, including earlier coats that have subsequently been covered, contains lead. Because of the problems associated with removing lead-based paint, be sure to review the earlier section on lead-based paints, as well as the section on lead-based paint in the "Rehabilitation" section. Because of the potential health problems and the cost of paint removal, new siding may be a better alternative.

Roofing

A number of types of roof coverings are encountered on older homes. Regardless of the type of roof, a leak should be obvious from damage inside the house. An inspection of the attic may also reveal water stains on the rafters, indicating small leaks that will eventually cause damage. Water damage inside the house may not be caused by faulty roofing; it may result from faulty flashing or condensation of moisture vapor within the house.

On built-up flat roofs, the sheathing may be inaccessible, but on other types of roofs where it is visible and accessible, its condition should be checked. Inspect the sheathing around roof edges, particularly adjacent to rain gutters. Rain runoff is frequently absorbed into the wood sheathing behind gutters. A small amount of deterioration from runoff is acceptable, but significant deterioration may require replacement. If general deterioration of the roof, as described in ensuing sections, indicates that reroofing is necessary, the deteriorated roof sheathing can be replaced at the same time.

Wood Shingles

Wood shingles are usually made of durable woods such as No. 1 or 2 grade cedar. A good wood shingle roof will last 20 years or more in favorable conditions.

As with shingles used for siding, a good wood shingle roof should appear as a perfect mosaic. Old wooden shingles eventually become porous, and can no longer keep out the rain. A wood shingle roof in poor condition has an overall ragged appearance with broken, warped, and upturned shingles. If a few shingles on a roof are cracked, split, or loose, they can be replaced. If the roof is generally deteriorated, however, it should be replaced even if there is no evidence of leakage.

Figure 17—Mildew on paint is most common in warm, humid climates as well as in shaded or protected areas.

Figure 18—Streaked water-soluble extractive discoloration can result from water wetting the back of one piece of siding and then running down on the front of the next piece.

Too much shade can foster moss growth, which promotes decay in wooden roofs by keeping them damp. Moss should be removed and the affected area treated with a wood preservative.

Asphalt Shingles

Asphalt and fiberglass-reinforced asphalt shingles are made in a wide range of weights and thicknesses. A good asphalt shingle roof should last 18 to 20 years.

The most obvious evidence of deteriorating asphalt shingles is loss of surface granules. Inspect the soil at the base of downspouts for accumulation of granules that have eroded from shingle surfaces.

Asphalt shingles can also become brittle with age. Inspect roof surfaces for missing or cracked shingles or shingles that show signs of cupping or bulging. Deterioration also occurs in the narrow grooves between the tabs or sections of the shingles or between two consecutive shingles in a row. This deterioration is not likely to be clearly visible from the ground, but inspection from a ladder or with binoculars should reveal it.

Certain types of asphalt shingles may contain asbestos. Their removal involves the hazards indicated under "Asbestos" in the "Condition Assessment" section.

Built-Up Roofing

Built-up roofing is used on flat roofs or on roofs with a low pitch, generally not more than 2-in-12 (2 in. vertical per 12 in. horizontal). The roofing usually consists of alternating layers of tar and roofing paper (also called roll roofing or felt roofing), sometimes covered on flat roofs with a layer of fine gravel or stone. The life of a built-up roof varies from 15 to 30 years, depending on the number of layers of roofing paper and the quality of application.

Built-up roofs should be examined for bubbles or blisters between layers, cracks in the covering, soft spots, puddles, and openings in seams between tar-covered strips of roofing paper. Such defects indicate the need for major repairs. A network of fine cracks on the roof surface (called alligatoring) is not necessarily an indication of serious problems.

Inspect the edges of the roof covering at the intersection of the roof with vertical structural elements such as chimneys, skylights, plumbing vent pipes, and exterior walls that extend above the roof line. Look for cracks in the roof covering and separation of the covering from vertical elements.

Other Types of Roofs

Other types of roofing materials encountered in older homes include metal, tile, and slate. All of them have a long life expectancy—60 years for a painted metal roof, 40 to 80 years for tile and slate.

Tin, terne, and other metal roofs are made by joining the edges of metal sheets together. Check metal roofs for joint tightness and corrosion. Defective areas should be patched and repainted.

Inspect tile and slate roofs for cracked, broken, or missing pieces. Leaks in these types of roofs are often corrected by simple patching, performed by persons experienced in working with these materials.

Flashing

Flashing consists of strips of metal or asphalt composition material that are used to seal the areas where the roof intersects walls, chimneys, or vents, and where two roof sections intersect to form a valley. Leakage frequently occurs at flashing points.

With a flashlight, examine the underside of the roof in the attic for stains or discoloration where the roof is penetrated by chimneys and vents and where the roof intersects a wall. Leaks around flashing can usually be sealed with cements made for the purpose, although such repairs may not last longer than about 5 years. If flashing is visible on the roof, check for corrosion; corroded flashing should be replaced.

Also, check for corroded gutters and downspouts, which can be restored by repainting unless they are severely corroded.

Roof Overhang

Certain houses, notably smaller models built shortly after World War II, were constructed without roof overhangs. Adding an overhang to the roofs of such houses should be considered as part of a rehabilitation plan; overhangs reduce maintenance and prolong the life of siding and window trim.

Soffit and Fascia	The soffit and fascia are elements of the cornice, which is the portion of the roof that overhangs the eave line. The fascia is a flat board, band, or face, used by itself or in conjunction with moldings, and located at the outer face of the cornice. The soffit board forms the underside of the cornice; it is often associated with a molding placed along the juncture of the soffit board and the siding. The soffit and fascia should be checked for decay and insect damage.
Chimneys and Fireplaces	Chimneys should be checked for cracks in the masonry, cracks in the chimney cap and flue lining, and deteriorated mortar (Fig. 19). The walls of the chimney should be at least 4 in. thick. Two flues in the same chimney structure should be separated by a solid masonry divider wall, normally made of 4-in. brick (Fig. 20).

A traditionally lined chimney has a flue liner made of sections of terra cotta, a type of baked clay. This liner should be at least 5/8 in. thick, and the sections are joined, one on top of another, with mortar.

In certain older chimneys, this liner is absent. A chimney without a flue liner should be regarded as unusable, since it does not meet modern safety standards. The only possible exceptions are unlined chimneys that are three courses of brick in thickness with sound mortar. Such a chimney should be inspected by local fire authorities to determine its safety.

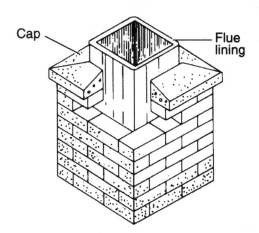

Figure 19—Parts of chimney showing flue lining and cap.

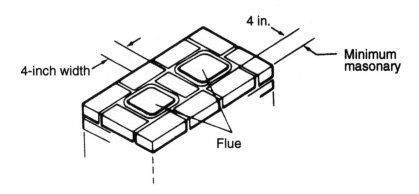

Figure 20—When two flues are in the same chimney structure, they should be separated by at least 4 in.

To see if the chimney is lined, or to check the condition of an existing liner, open the damper and look up the chimney using a powerful flashlight. The chimney liner should begin above the small portion of the chimney where the bricks proceed upward at an angle from the damper. The liner is distinguishable by the absence of brick mortar joints.

Some old chimneys were constructed with a partial liner, located only at the top and bottom. Such a chimney should not be regarded as safe. To be certain that a chimney is fully lined, have a chimney sweep or brick mason check the full chimney by lowering a drop light down the chimney from the top.

Inserting a terra cotta liner into an old chimney is a feasible but difficult project, which involves breaking into the chimney from inside the house and inserting the terra cotta sections. Another method for retrofitting a chimney with lining involves inserting a stainless steel pipe and packing vermiculite insulation between the pipe and the inside surface of the chimney. This is the preferred lining if the chimney will be serving a woodstove.

A third method involves pouring a cement slurry around the outside of an inflatable cylindrical form that is lowered into the chimney and centered by use of spacers. This procedure strengthens the chimney as well as provides a lining.

Check in the attic to ensure that the ceiling and roof framing are no closer than 2 in. to the chimney. Failure to meet this requirement constitutes a fire hazard that should be corrected.

To meet the requirements of the major model fire codes, the top of the chimney must extend vertically a distance of 3 ft above the point where it penetrates the roof, and the top of the chimney must be at least 2 ft above any portion of the house or roof that lies within a 10-ft radius around the chimney (Fig. 21).

Rain caps and animal screens are recommended for the tops of chimneys, although they are not required by fire codes.

If the house has a fireplace, check that it has an operating damper. Where there is no damper, add one to prevent heat loss up the flue when the fireplace is not in use. The condition of the fire brick in the fireplace and the condition of the hearth should be checked. Glass fireplace doors can increase the safety of the fireplace.

A fireplace that has been used substantially probably draws well. The draw can be checked by lighting a few sheets of newspaper in the fireplace. A good fireplace will draw immediately; a usable one will draw after about a minute.

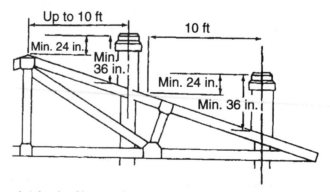

Figure 21—Proper height of a chimney relative to its height and location on roof.

Exterior Utilities
Examine the condition of the electrical service entrance, and verify that there is an airtight seal on this and other utility entrances.

Check if there are outside light fixtures at each exterior door and inside the garage or carport and whether they are operable. Lights in the garage or carport should be switched both at an exterior entrance and inside the house.

Determine if weatherproof GFCI duplex receptacles exist and are operable in garage/carport and patio/balcony areas. (GFCI stands for ground fault circuit interrupter, a device that cuts the power to an outlet if current from that outlet is grounded by human contact.)

Open external hose bibs to verify that they operate and that water pressure is adequate. Adequate pressure with a fully opened bib should fill a quart container in 5 seconds or less.

Porches and Decks
Open porches are particularly susceptible to decay and insect attack. Decay is especially likely to occur at joints, particularly end-grain to end-grain and end-grain to side-grain joints. Water can easily penetrate the joint, where it is absorbed readily by the end grain of the wood, increasing the moisture content in that area. Lack of air flow within the joint makes removal of excess moisture difficult or impossible, and higher moisture content increases the hazard of decay. Railings can become loose as a result of wood decay, corrosion of the fastenings, or mechanical damage, and thus become a hazard to the user. Stringers of stairs leading from the porch to the ground are particularly susceptible to decay if they rest on the ground and are not treated with a wood preservative.

Check the condition of wood members, particularly where they are joined. Inspect the porch roof and its supports; the roof edge should be parallel with the edge of the house roof, and attachments to supports should be firm. If braces are present where the roof joins the house, check for a firm connection to the house and roof frame. The bottoms of porch columns or other supports are particularly susceptible to decay and should be checked carefully.

Decks and patios have many of the same problems as porches. Joints are particularly susceptible to decay because they trap and retain moisture. Check for decay where the deck is attached to the house, as well as where any supporting members or steps meet the soil. Check the fasteners (nails, bolts, etc.) to make certain they have not corroded and that the joints are still tight.

Basement

Moisture
In addition to problems relating to foundation settlement (see "Foundation" in the "Condition Assessment" section), damp or leaky basement walls may require major repairs, especially if the basement space is to be used as a living area. Basement dampness can be caused by cracks in walls, clogged drain tiles, clogged or broken downspouts, inadequate slope of the finished grade away from the house foundation, or a high water table. Check for dampness by examining the basement and the sump a few hours after a heavy rain.

The most common source of dampness is surface water, such as from downspouts discharging directly at the foundation wall or surface drainage flowing directly against it. Determine if surface water accumulates during a heavy rainstorm and how it can best be diverted. The cardinal rule is to keep water away from the foundation, which is best accomplished by proper grading.

A high water table is a more serious problem. There is little possibility of achieving a dry basement if the water table is high or periodically high. Heavy foundation waterproofing or footing drains may help, but since the source cannot be controlled, these techniques will do no more than minimize the problem.

Other Problems
One or more of the beams that support the floor joists may be sagging along its length. One method of solving the problem is to jack up and support the beam where it sags the most. A simple solution is to place an adjustable steel column beneath each low point. Such columns have built-in screw jacks that ensure proper support for the beam.

Check the condition of the overhead floor framing. Water stains on the underside of the overhead floor may indicate a regular leak somewhere above. Check floor drains for odor and slow or clogged operation.

Check the squareness and condition of the exterior entrance door, window sills, jambs, and headers, and the condition and drainage of the exterior entrance and window area. Note any decay or insect damage in windows, window sills, joists, and beams. Windows should be operable.

Check the soundness of the basement stairs. In some old houses, basement stairs may be narrow and steep even if they are in good condition. If space permits, consider replacing them for safety.

Basements are the primary entry point for radon. Methods for determining levels of radon in the home have been discussed earlier under "Radon" in the "Condition Assessment" section.

For energy conservation, check if insulation has been placed on the inside of band or header joists (Fig. 22); lack of insulation in these areas can cause significant heat loss.

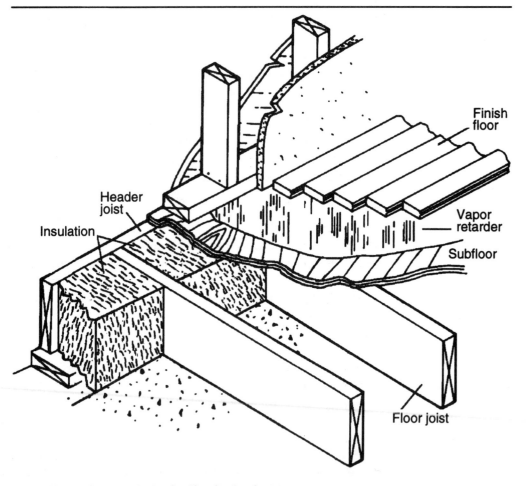

Figure 22—Insulation on the inside of band or header joists.

Living Areas

Living areas discussed here include all areas of the house other than the basement, kitchen, bathrooms, and attic, which are treated separately.

Wood Floors

In examining wood floors, look for buckling or cupping of boards, which can be caused by high moisture content in the boards or wetting of the floor. Look for boards that are separated because of shrinkage; shrinkage is more probable if the boards are wide. Separation between boards generally has no adverse affect on structural strength, but can pose problems for cleaning.

If the floor is generally smooth and without excessive separation between boards, refinishing may restore it to good condition. However, before sanding, be sure there is enough thickness left in the flooring to permit sanding. Most flooring cannot be sanded more than two or three times; if it is softwood flooring and no subfloor is present, even one sanding might weaken the floor too much. Sanding plywood block floors is not recommended.

If floors are too thin to sand, some type of new flooring will have to be added. Adding new flooring is also an option if there are wide cracks between the boards of the existing floor. The owner may wish to retain old wideboard floors for the sake of their appearance, regardless of the problems involved in keeping them clean.

Tile and Sheet Flooring

Resilient tile and sheet flooring are water resistant and commonly used in kitchens, bathrooms, and other work areas. However, they may appear anywhere in a house. Common defects in these materials include loose, missing, damaged, badly worn, or warped tile and torn sheet flooring. If a single tile must be replaced, the flooring in the entire room may have to be replaced since the tiles change color and tonality with age and wear, and replacement tiles may not provide a satisfactory color match.

Underlayment for resilient tile floors and sheet flooring usually consists of plywood, hardboard, or particleboard placed on top of the subfloor. Problems caused by the underlayment appear as straight-line cracks, ridges, or grooves in tiles or sheet flooring. Raised nailheads can also show through the flooring; this is caused when the joist into which the nail has been driven flexes through the underlayment and sheathing or by moisture problems. If this happens, the underlayment and the damaged flooring should be replaced. Raised nails should be removed while the underlayment is exposed.

If there is no practical alternative to tearing up the old floor, the procedures described in the publication *Recommended Work Procedures for Resilient Floor Coverings*, issued by the Resilient Floor Covering Institute and revised in 1987, should be followed carefully. Rules governing the handling of asbestos that were issued by the Occupational Safety and Health Administration (OSHA), U.S. Department of Labor, in September 1988, state, "There appears to be virtually no possibility that the excursion limit (defined as one asbestos fiber per cubic centimeter of air, averaged over 30 min) would be exceeded if the recommendations of the Resilient Floor Covering Institute were followed."

Ceramic or quarry tile floors may have cracked, loose, or missing tiles, and grout may be deteriorated or missing. These, of course, will be nearly impossible to replace because of the difficulty in obtaining matching material. New grout can be applied.

Walls and Ceilings

The interior wall covering in old houses usually consists of plaster. Gypsum wallboard is found in newer homes and in older homes that have been renovated or remodeled. Wood paneling may also be found; it could be wainscoting and may be confined to a single room or area.

Swollen or crumbling plaster or gypsum board indicates that moisture is present; the source should be found and eliminated.

In houses built before the mid-1930s, plaster walls and ceilings were installed by first nailing thin strips of lathing, placed close together, to the house framing. Three coats of plaster were then applied over the laths. The plaster filled the open spaces between the laths and spread out behind the laths, forming long "keys" that anchored the plaster.

The plaster walls and ceilings of wood lath houses are heavy. In the ceilings, the keys may deteriorate, posing the risk that sections of the ceiling may fall, possibly injuring anyone who may be directly beneath. Deterioration of the keys in ceiling plaster can be difficult to detect until a portion of the ceiling actually collapses. Apparent looseness or sag in any area of the ceiling may indicate weakness. Such areas can be tested by upward pressure with a broomstick or directly with the hand.

In houses built after the mid-1930s, perforated gypsum board was used in place of wood lath as an underlayment for plaster. Thinner coats of plaster were applied, resulting in lighter walls and ceilings. In these "rock lath" houses, the plaster walls are stronger than the plaster walls in wood lath houses, and the danger of loose or falling ceiling plaster is greatly reduced.

Old plaster frequently has cracks. Although the homeowner may wish to repair them for cosmetic purposes, they are not usually a structural problem. One or more lengthwise cracks in the ceilings of a wood lath house should not cause alarm if the plaster ceiling is not loose. Cracks are also common where walls meet and where walls and ceilings meet.

A crack that should not be ignored is one that runs continuously across part of the ceiling and an adjacent wall. Such a crack may indicate that the foundation has settled, which should be checked by a competent professional engineer.

In more recent construction or in renovated older structures, walls and ceilings are usually made of gypsum board that has been painted. Common defects include nail pops—protruding nailheads—and faded or discolored paint. Such defects should be corrected.

Interior Coatings

The paint on interior surfaces may be excessively thick. It may be chipped because of mechanical forces, or it may be peeling because of incompatible successive coats or because the surface was improperly prepared before it was repainted.

As discussed earlier (see "Lead" in the "Condition Assessment" section), lead-based paint was widely used in residential applications for both interior and exterior surfaces until the early 1940s. Assume that interior paint in dwellings built prior to that time, or at least the earlier coats on surfaces that have been repainted a number of times, contains lead. There are health hazards involved in attempting to remove it, and these must be considered in any rehabilitation (see "Lead" in the "Rehabilitation" section).

Stairs

Common defects of old staircases include deteriorated finish; worn, loose, uneven, warped, or broken treads; deteriorated stringers; and deteriorated or loose railings.

In certain instances, old stairs may be narrow and steep, inviting accidents. Basic dimensions for modern stairway design include headroom clearance, stairway width, stair tread run, and stair rise. Generally accepted dimensions as shown in Figure 23 are

- minimum headroom clearance of 6 ft 8 in.,

- minimum stairway width of 36 in.,

- minimum stair tread run of 9 in., and

- maximum stair rise of 8-1/4 in.

The extensive structural changes that may be required to renovate stairs could preclude replacing steep or narrow stairs while they remain sound; railings, however, may have to be added if they are not already present. If extensive interior renovation is being considered, then it may be feasible to redesign the stairway.

Attic

As is indicated in the earlier section on roofs, the underside of the roof in the attic should be checked for stains or discoloration, particularly where the roof is penetrated by chimneys and vents and where it intersects a wall. Leaking is most common at these points.

Ventilation is required in most attics to help remove water vapor and condensation. This need is more fully appreciated today than it was when many of America's older homes were built. The problem of adequate attic ventilation should be addressed as part of the total rehabilitation assessment of an older dwelling.

During cold weather, the warm, moist air from heated rooms can work its way into the attic space around the places where pipes and electrical fixtures penetrate walls and ceilings and through other inadequately protected areas. In colder climates, vapor retarders are commonly installed in the ceilings below attics in modern home construction to help prevent significant moisture migration through the ceiling. These vapor retarders are generally not found in unrenovated older dwellings.

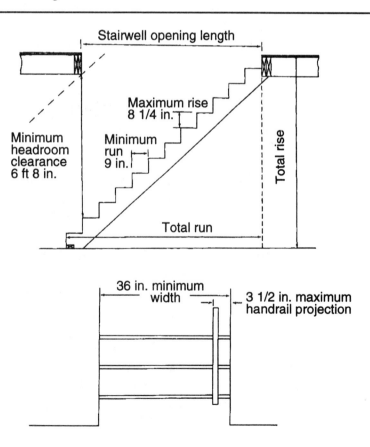

Figure 23—Stairway design requirements and terminology.

Wood shingle and wood shake roofs do not resist the movement of vapor from the attic to the outdoors. Roofing materials such as asphalt shingles and built-up roofs, however, are highly resistant to moisture movement, which can contribute to the buildup of water vapor in the attic.

During winter weather, an inadequately ventilated attic becomes warm and can foster the formation of ice dams at cornices or in roof valleys. The warm air in the attic can warm the roof sheathing, causing snow on the roof to melt. The water runs down the roof and refreezes on top of the cold overhang (Fig. 7). The ice dam forms a trough that retains water at a point just above the house wall. The water soaks under the shingles and sheathing and can damage interior and exterior paint, drywall, and structural wood. Inadequate ventilation can also cause moist, warm air to condense on the underside of the roof sheathing; this creates potential decay and could render insulation ineffective if water drips onto it.

If a vapor retarder has been used in the ceiling below the attic, 1 ft^2 of net free-ventilator area is recommended for each 300 ft^2 of ceiling area. The term net free-ventilator area refers to the gross vent area minus the area of obstructions such as screens and louvers (for conversion factors, see Table 4).

Without a ceiling vapor retarder, the recommended net free area should be doubled. However, it need not be doubled if at least half the net free area is located in the upper portion of the space 3 ft or more above the eave and the rest of the venting is provided by eave or soffit vents.

Kitchen

The kitchen is an important and heavily used area of the home. New appliances and new concepts of convenience have brought about great changes in the layout, appearance, and functioning of kitchens. In an older house, replacing old appliances and fixtures and substantial remodeling may be a priority, even if the existing kitchen is basically functional.

Common defects in kitchens include inadequate space for movement, poor overall arrangement, inadequate work surfaces, outdated appliances, inadequate storage space, inadequate ventilation and lighting, inadequate number and placement of electrical receptacles, and inconvenient access to the dining area.

As discussed in "Return on Investment" in the Introduction, professional rehabilitation of the kitchen offers an average 73 percent return on investment, one of the highest return percentages for various home rehabilitation projects. This reflects a clear buyer preference for modern kitchens, even if the house is not new.

Layouts

Basic movements in food preparation are from the refrigerator to the sink and then to the range. Four generally recognized types of layouts are the U and L, the corridor, and the side wall (Fig. 24). The work triangle is smallest in the U and corridor layouts. The side wall arrangement is preferred where space is limited, and the L arrangement is used in a relatively square kitchen that must accommodate a dining table.

Table 4—Multipliers for various vent coverings to determine net free-vent area

Type of covering	Area of opening (x required net free area)
1/4-in. hardware cloth	1
1/4-in. hardware cloth and rain louvers	2
1/8-in. mesh screen	1.25
1/8-in. mesh screen and rain louvers	2.25
1/16-in. mesh screen	2
1/16-in. mesh screen and rain louvers	3

In a layout that has become increasingly popular, the kitchen area is substantially open to an adjoining living, living–dining, or family area. In some of these layouts, the stove, with accompanying floor cabinets and workspace or eating counter, constitutes an "island" between the rest of the kitchen and the adjoining area (Fig. 25).

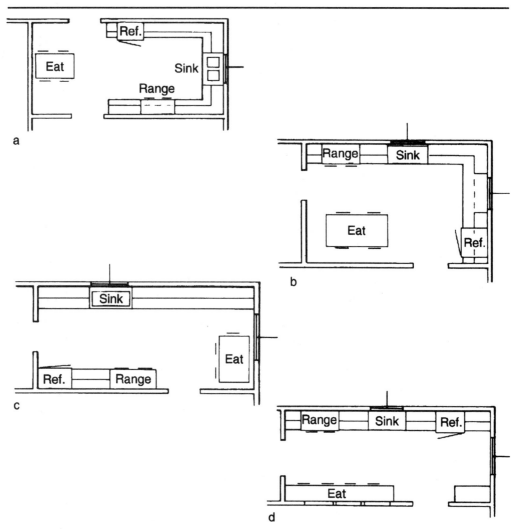

Figure 24—Kitchen arrangements: (a) U-type, (b) L-type, (c) corridor, and (d) side wall.

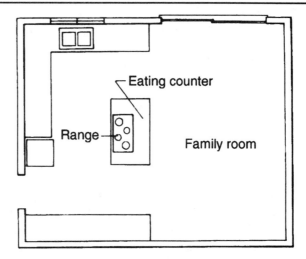

Figure 25—Island counter dividing kitchen and family room.

The kitchen should be designed so there is no "through traffic" to adjoining rooms. Entryways in corners should be avoided. Where entryways to the kitchen have doors, the swings should not obstruct appliances, cabinets, or other doors. If swinging the door places it in the path of travel in a hall or other area, consider installing a sliding or folding door.

If extensive remodeling is not undertaken and if kitchen cabinet space is adequate and arranged well, updating the cabinets may be desirable. New doors and drawer fronts can be added to the old cabinet framing. Refinishing old cabinets and installing new hardware can improve an old kitchen.

New countertops can also improve a kitchen's appearance. Tops can be purchased at lumber supply stores, or they can be fabricated and installed by a custom counter shop. Plastic laminate over particleboard backing is commonly used.

There should be adequate windows to make the kitchen light and cheerful. Current trends toward indoor–outdoor living have fostered the patio kitchen, with large windows over a counter that extends to the outside to provide an outdoor eating area. This arrangement is particularly useful in warm climates, but is also convenient for summer use in any climate.

Placing the sink on an outside wall makes it possible to look out the window while working at the sink. This pleasant arrangement can be more costly than installing the sink along an interior plumbing partition because of the possibility of freezing pipes and the additional costs to prevent it.

Much research in kitchen planning has been done by the USDA's Agricultural Research Service and by various state universities. Contact the Agricultural Extension Service for bulletins on kitchen planning from these sources. Many companies that build kitchen cabinets will also assist in planning.

Appliances

In deciding where to place appliances, keep in mind the following list of average life expectancies:

Appliance	Average life expectancy (years)
Range	15–16
Refrigerator	13
Dishwasher	11
Garbage disposal	10

Recommended refrigerator sizes are as follows:

Number of bedrooms	Refrigerator size (ft³)
2	14
3	17
4 and 5	19

Floors

Kitchen floors are generally covered with a water-resistant surface, usually vinyl tile or vinyl sheet flooring. Common defects of older floors include missing, damaged, broken, or warped vinyl tile; torn vinyl sheet flooring; swollen or deteriorated underlayment caused by water leakage under the floor covering; nail pops or seams showing through the floor covering; and worn or discolored flooring.

Bathrooms

Old houses often lack adequate bathroom facilities by contemporary standards. Many have only one bathroom, and it is likely to have old, stained fixtures and no shower. An additional half-bath or a second bath should be considered for houses with only one bathroom.

As with kitchen remodeling, improvements of existing bathrooms offer one of the highest returns on investment among home rehabilitation projects (see "Budgeting for Rehabilitation"). The figures quoted in "Budgeting for Rehabilitation" do not necessarily pertain to adding a new bath or half-bath, which involves both structural and plumbing modification to the home, but relate to modernizing the existing bathroom. The extent of these modifications depends on several factors that are discussed later. However, there can be little doubt that such modifications are a desired feature of modern living.

Finding a convenient location for an additional bath can be difficult. Adding a room onto the house is seldom a good solution, since it is usually desirable to have access to the bath from a bedroom hallway. A half-bath, however, can be satisfactorily located near the house entrance or in a work area.

A significant economic consideration in locating a new bath is to keep all plumbing runs as short as possible, preferably on a single wall. In addition, fixtures on one plumbing wall can use a common vent. Bathrooms built on both floors of a two-story house are most economically built with the second floor bath directly above the first floor bath.

A mistake that is sometimes made in adding a bath to an older house is to place it in any unused space without regard to convenience. Many bathrooms have been placed in what was formerly a pantry or a large closet, or under a stairway. The bathroom can then be accessed only through the kitchen or bedroom or is totally removed from the bedroom area. If this mistake has been made in the house being assessed, it may be desirable to add another bath in a better location.

In a house with large bedrooms, a portion of one bedroom might be taken for a bath (Fig. 26). The bedroom should be at least 16 by 10 ft after the bath is built.

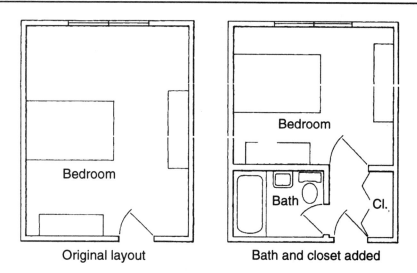

Original layout · Bath and closet added

Figure 26—Portion of large bedroom used to add bath.

If the bedrooms are small and all are needed, an addition may be the only alternative. Converting a small bedroom into a bath (or two baths) and adding another bedroom may be advantageous. For second-floor bedrooms, the wall containing the plumbing must have a wall directly below it on the first floor, through which the plumbing can pass.

In a one-and-a-half story house, it may be possible to add a bath in the shed dormer (Fig. 27). As with two-story houses, the wall containing the plumbing must have a wall directly below it on the first floor.

The minimum size for a bathroom is 5 by 7 ft (Fig. 28), but larger sizes are desirable. A larger bathroom is less crowded, and has more space for towels, cleaning equipment, or supplies (Fig. 29).

If the plan calls for only one bath, it may be desirable to divide it into compartments for use by more than one person at a time (Fig. 30). Two full baths can be economically built with fixtures back to back (Fig. 31), but a convenient location should not be sacrificed for this.

Bathroom fixtures vary in size, and fixture dimensions from building supply dealers are necessary to make detailed plans. Minimum dimensions between and around fixtures are shown in Figure 32.

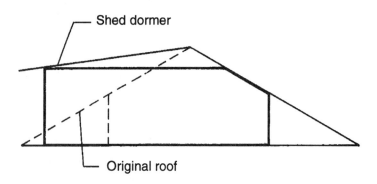

Figure 27—Shed dormer for additional space.

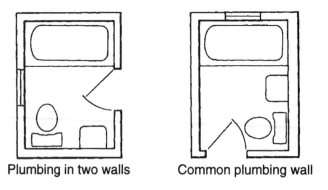

Plumbing in two walls   Common plumbing wall

Figure 28—Small-size bathroom (5 by 7 ft).

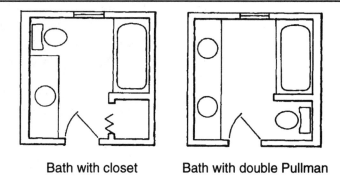

Bath with closet   Bath with double Pullman

Figure 29—Moderate-size bathroom (8 by 8 ft).

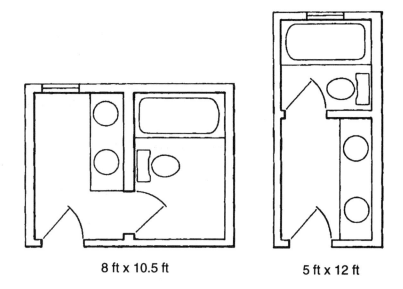

8 ft x 10.5 ft   5 ft x 12 ft

Figure 30—Compartmented bathroom.

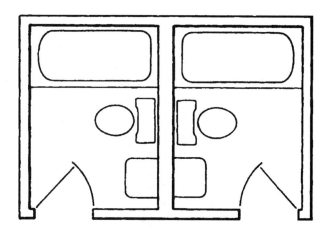

Figure 31—Two bathrooms with economical back-to-back arrangement.

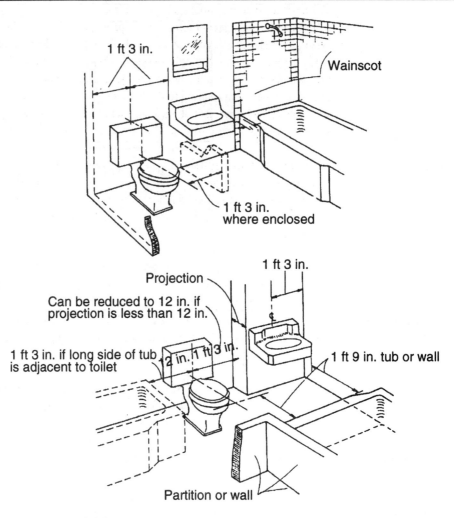

Figure 32—Recommended dimensions for fixture spacing.

Fixtures

Existing bathrooms can often be conveniently and attractively modernized by the use of prefabricated and presized tub liners, shower surrounds, window surrounds, and modern hardware that can be installed without removing existing fixtures or modifying existing piping runs.

For purposes of both water and energy conservation, the homeowner should consider installing shower heads with flow control devices that limit water flow to a maximum of 2.5 gal/min at pressures between 20 and 60 lb/in^2. If toilets are being replaced, install water-saving models.

Floors

Bathroom floors, like kitchen floors, are often covered with vinyl tile or vinyl floor covering that contains asbestos. Potential problems with these floor coverings have been discussed previously. The condition of other types of tile needs to be assessed. Ceramic tile, particularly in mosaic form, is common in bathrooms of older houses.

Utility Assessment

The basic utility systems in the home are the heating, ventilation, and air-conditioning equipment (HVAC); plumbing; and electrical distribution. These systems are partially concealed, and their adequacy may be difficult to determine. For the same reason, major changes to HVAC and plumbing systems involve cutting wall surfaces. Nevertheless, in an old house, major changes in one or more of these systems or total replacement may be necessary, which will be a major expense.

The existing systems should be checked for safety and for conformity to the local code. In many jurisdictions, local building code officials will inspect heating, plumbing, and electrical systems for safety and code conformity, and their services should be utilized. The homeowner may also wish to have the utility systems checked by knowledgeable professionals. This professional review can include an assessment of the condition and operability of the systems, and their capacity in relation to the homeowner's requirements. For most homeowners, it is wise to rely on building code officials and professionals for detailed evaluation of home utility systems. The ensuing information provides general guidance for evaluation.

Heating

Advances in heating systems and concepts of comfort have outdated the heating systems that constituted original equipment, and sometimes even replacement equipment, in old houses. Central heating is considered a necessity in all but very small houses. Even if other older types of heating systems are still operational, updating them could provide increased comfort and energy savings.

The best way to check the adequacy of the heating system is to use it. If the system is adequate for the desired degree of comfort, check the overall condition of the furnace or boiler.

Gravity Warm-Air Systems

This type of system is common in older homes. Some gravity warm-air furnaces may heat a house relatively well, but the temperature control and heat distribution will probably not be as good as with a forced-circulation system.

If a warm-air system is exceptionally dirty, there may be smudges above the registers. This will indicate that some repair work is required; if the furnace is old, it may need to be replaced. Rusty ducts may need replacing also.

Floor Furnace Systems

Warm-air floor furnaces may be adequate for small houses, but their condition and adequacy should be assessed.

Gravity Steam Systems

One-pipe gravity steam-heating systems, which are similar in appearance to hot water systems, are common in older homes. If properly installed, they will provide adequate heat but with no great speed or control. They can be modernized without basic changes by replacing standing radiators with baseboard convectors.

A one-pipe gravity steam system can be made more positive in action by conversion to a two-pipe system. This conversion involves the addition of traps and return lines.

A two-pipe steam system can be modified to a circulating hot water system by adding circulating pumps. This results in faster heat distribution and excellent heat control.

Radiant Heat Systems

Radiant heat from hot water flowing through coils embedded in concrete floors or plastered ceilings is not common but may provide excellent heating. Breaks in ceiling coils can be repaired, but repairing breaks in floor coils is extremely difficult. If floor coil breaks are extensive, the system probably will have to be replaced.

Electric Panels

Electric heating panels have no moving parts to wear out and should be in good operating condition unless a heating element has burned out. The ability of the system to provide adequate heating should be checked.

Air-Conditioning

Modern forced-air heating systems use ductwork to distribute heated air throughout the house. These ducts can also be used to distribute cool air from central air-conditioning units. If a house does not have a duct system, adding it for the purpose of providing central air-conditioning can be very expensive. Individual air-conditioning units mounted in windows or installed in walls offer a less expensive alternative.

In older homes that have been retrofitted with central air-conditioning, the homeowner should consider having a professional engineer or qualified contractor evaluate the capacity of the system in relation to the size of the space that it must serve.

Oversized air-conditioning units can create undesirable humidity levels in the home. In addition to cooling, air-conditioning dehumidifies the indoor air. As air comes in contact with the cooling coil in the air-conditioner, moisture in the air condenses and is drained away. The air that is redistributed to the house is both cooler and drier. However, if the air-conditioning equipment is oversized, it will cool the air and shut off before it dehumidifies the air adequately. The result can be both decreased physical comfort and long-term problems with the house that are a result of excessive humidity. This problem can be particularly important in warm, humid climates such as those in the Gulf Coast States region of the United States.

Water System

Water Supply

In modernizing an old house, it may be necessary or desirable to add water supply and drain lines. New lines may be required for automatic washers, for added bathrooms, or for outfitting a reorganized layout.

Many old homes have lead pipes in their water system, which can constitute a major health hazard. See previous sections for information on this problem.

Water pressure is important. Check several faucets to see if the flow is adequate. Low pressure can be the result of a service that is too small or reduced because of lime deposits or corrosion in galvanized pipes.

The main distribution pipes should have 3/4-in. inside diameters, but branch lines may have 1/2-in. inside diameters. Sizes can be checked easily. Copper pipes with 1/2-in. inside diameters have 5/8-in. outside diameters, and pipes with 3/4-in. inside diameters have 7/8-in. outside diameters. Galvanized pipes with 1/2-in. inside diameters have 7/8-in. outside diameters, and pipes with 3/4-in. inside diameters have 1-1/8-in. outside diameters.

The supply pressure may be inadequate. If the house has its own water system, check the gauge on the pressure tank; it should read a minimum of 20 lb/in^2 and preferably 40 to 50 lb/in^2. Anything less will indicate that the pump is not operating properly or the pressure setting is too low. If the supply is from a municipal system, the pressure in the mains may be too low, although this is unlikely.

Check shutoff valves at the service entrance and at various points in the system to determine if they have become frozen with age or disuse. If the system includes a water meter, it should register zero flow when the valve is closed.

The location of supply lines should be checked to determine if portions of the lines are susceptible to freezing in unheated areas and against or in exterior walls.

Check for leaks in the water supply system. These are indicated by rust or dampness at pipe joints.

Water hammer may be a problem. This banging noise results from stopping the water flow in a pipe by abruptly closing the faucet. Air chambers placed on the supply lines at the fixtures usually absorb the shock and prevent water hammer. If there is water hammer, existing air chambers may be waterlogged. Air chambers can be added, if necessary.

Water from a private well should be tested for quality, whether or not the well has been in continuous use.

Water Heating

A hot-water heating system can also be used to heat water for bathing and other household needs. However, in a hot-air furnace, the water-heating coil seldom provides enough hot water. In addition, during summer months when heating is not needed, a separate system is required to provide hot water. A gas water heater should have at least a 30-gal capacity, preferably more. An electric water heater should have a capacity of 50 gal or more.

Fixtures

Old fixtures may be rust-stained and discolored, making replacement desirable, or the home-owner may wish to replace them with newer and more contemporary styles that require less maintenance. Replacing washers is an example of this. In general, installing tasteful contemporary kitchens and bathrooms enhances rather than detracts from the value of an old house in which period features are otherwise extensively retained.

Drainage

The drainage system consists of the sewer lateral, the underfloor drains, the drainage pipes above the floor, and the vents. Pipes may be clogged or broken, or they may be too small. Venting may be absent or inadequate and below code requirements.

Flush fixtures to see if drains are sluggish. If so, these are some possible sources of the problem:

- Old laterals are commonly made of vitreous bell tile. They may have been poorly installed or may be broken. The problem can be compounded by tree roots entering at the breaks or through the joints. Roots can be removed, but this may have to be done every few years.

- The underfloor drains may be made of tile or steel. They may be broken or rusted out, or they may be clogged and only need cleaning.

- The drainage system above the basement floor may be too small or may leak.

- Vents may be inadequate or clogged. In extreme cases, this may cause the water in the traps to be siphoned out, allowing sewer gas to enter the house. Note any excessive suction when a toilet is flushed.

Sewage Disposal

Run water for a few minutes to check for clogged drain lines between the house and the sewer main. If there is a private sewage system, the adequacy of the drain field should be checked by an official of the local health department or a qualified contractor, or both. If a new drain field is needed, some codes require that the soil be tested for percolaton.

Electrical System

Many new electrical appliances have come into common use in recent years, and old houses may not have adequate wiring to accommodate them, particularly if air-conditioning is installed. Lighting in the modern home also usually exceeds that which was deemed acceptable in the early 20th century. This will probably necessitate improvements in the electrical system.

Capacity	Electrical service should have a capacity of at least 100 amps (A) for the average three-bedroom house. If the house is large or if air-conditioning is added, the service should be 200 A. The capacity appears on the main fuse of the system with a fuse box. With circuit breakers, the capacity is stamped on the breaker labeled as the main. Additional circuits can be added to the main distribution panel if there is room; otherwise another distribution panel can be installed.
Distribution System	Examine electrical wiring wherever possible. Many times, wiring is exposed in the attic or basement. Wiring should also be checked at several wall receptacles or fixtures. If any armored cable or conduit is badly rusted, or if wiring or cable insulation is deteriorated, damaged, brittle, or crumbly, it has probably deteriorated from age or overloading and should be replaced.
	It is a good idea to have at least one electrical outlet on each wall of a room and two or more on long walls. Ceiling lights should be controlled by a wall switch, and rooms without ceiling lights should have a wall switch for controlling at least one outlet.
Electrical Safety Features	Electrical safety in the home is important. Two innovations in home electrical systems—the third-wire ground and GFCIs—have been adopted into the National Electric Code recently to reduce electrical hazards.
	The third-wire ground involves a three-pronged plug and receptacle; the third prong provides grounding for the equipment or device that is plugged in. In general, such plugs and receptacles are required for laundry areas, breezeways, garages, exterior outlets, areas around kitchen sinks, basements, workshops, and open porches. Most appliances and hand-operated tools require grounding.
	The GFCIs shut off power to a circuit when current is being diverted from the circuit directly to the ground by a conductor such as a human body. The amount of current required to trip a GFCI is much less than that required to trip a circuit breaker or blow a fuse.
	In today's codes, GFCIs are required around swimming pools and in both outdoor and bathroom receptacles. In addition, GFCIs are required for at least one receptacle in the basement and all receptacles in the kitchen that are within 6 ft of the sink and higher than the countertop. These requirements pertain to new construction. If third-wire ground receptacles or GFCIs are not present in older homes, it is a good idea to install them in new construction. The possibility of installing three-pronged plugs can be discussed with an electrical contractor.
Recognition of Damage	In most homes, decay is not a major problem. However, because of poor maintenance, improper construction, design errors, poorly built additions, or other major changes in the house, moisture may have accumulated, causing decay. Because damage caused by wood decay often is more prevalent than insect damage, take care to determine its presence, its extent, and the proper remedial action. Excess moisture and poor construction practices also encourage insect infestations. However, well-constructed houses can also become infested. Chemicals are normally used to prevent attack by subterranean termites and other insects. As with all pesticides, the chemicals should be applied in accordance with EPA requirements. Only licensed professionals can apply some chemicals.
Decay	Wood decay is caused by fungi, which are nongreen, flowerless organisms that use wood and other plant materials as a food source. The wood becomes infected with spores, which are the reproductive mechanisms of the fungi. Tiny threadlike growths called hyphae germinate from the spores. The hyphae secrete enzymes that attack the wood cells, causing the wood to disintegrate.

Fungi have four requirements for growth: a food source (wood), oxygen, a favorable temperature, and moisture.

The optimum temperature for wood decay fungi is 75°F to 90°F. Decay continues at a reduced rate, or is even dormant, at freezing or extremely high temperatures.

For fungi to destroy wood, free water must be present in the cell cavities. Fungi find free water when the moisture content of wood is 25 to 30 percent or higher. Fungal activity ceases at moisture contents of 20 percent or less. Thus, one effective way to control wood decay is to keep the moisture content of wood below 20 percent. Lumber properly installed in buildings should have a moisture content of 8 to 15 percent, whereas lumber cut from freshly harvested trees is often 80 percent or higher, based on the ovendry weight of the wood. Wood in contact with soil almost always has a moisture content of 20 percent or more, and thus is subject to decay.

The moisture content of lumber can be determined by use of a moisture meter (Fig. 33). The needles of the meter should be inserted in the side grain parallel to the grain of the wood. Moisture meters are accurate up to about 25 or 30 percent moisture content.

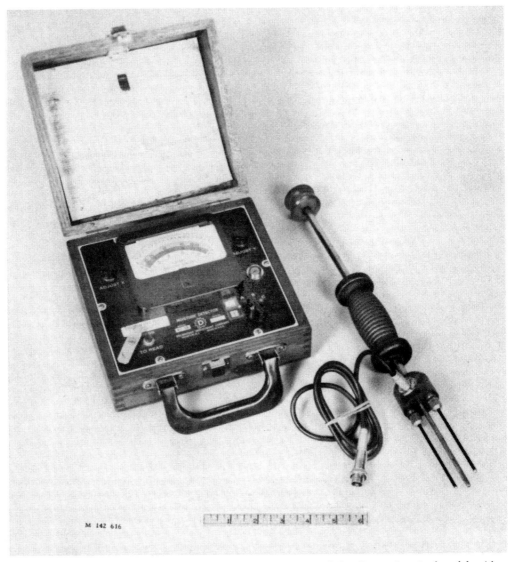

M 142 616

Figure 33—Typical meter for determining moisture content of wood. Smaller, pocket-sized models with short probes (needles) are also available.

Another way to control wood decay is to alter the wood chemically by applying a preservative. Wood that has been pressure-treated with a preservative chemical is available at retail lumber yards.

NOTE: Unused pieces of pressure-treated wood should be buried or sent to a landfill rather than burned. Burning pressure-treated wood produces toxic fumes and ashes.

Signs of Decay

Brown rot (Fig. 34) is the predominant type of decay in softwoods. In the early stages of decay, the wood surface lacks luster and appears dull or dead. As decay progresses, the wood acquires an abnormal brown color, as though it had been charred. Cross-grain cracking, collapse, or crumbling, and abnormal shrinkage finally result. The strength properties of wood attacked by brown rot fungi decrease rapidly, even in the early stages of decay.

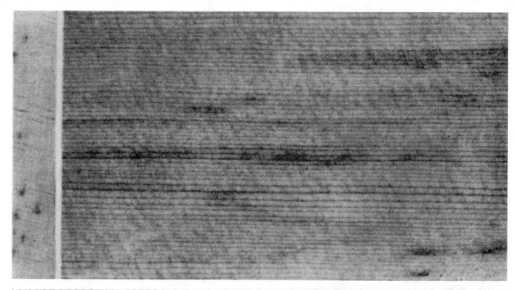

Figure 34—Early stage of brown rot showing discoloration of side grain (top right) and end grain (top left) of the same board. Cross-grain cracking and collapsed wood are associated with advanced decay.

Dry rot is an erroneous term used to describe decomposed, brown-rotted wood. It is used to identify wood that is both rotted and relatively dry—thus dry rotted. Actually, the wood was wet when it decayed and subsequently dried out.

White rot (Fig. 35) can appear in hardwoods used above ground. In the early stages, the wood color tends to turn off-white, sometimes making the wood appear bleached. Black zone lines may develop in the light areas. Unless severely degraded, the wood does not crack across the grain and does not collapse or shrink abnormally as with brown rots. A white fibrous mass may result.

In its early or incipient stage, wood decay is difficult to detect by visual inspection even though serious strength loss in the wood may have already occurred. The pick test, which includes lifting pieces of wood from the surface with a pointed tool (Fig. 36), can detect decay at a weight loss in the wood of approximately 10 percent or more. Serious loss of strength would have occurred at that weight loss.

Advanced decay is readily detected. Wood with appreciable decay will often break brashly and abruptly across the grain, whereas sound wood often splinters at the break. Brashness, reflecting reduced toughness, can be detected by breaking small pieces by hand or by the pick test (Fig. 36).

Wood can also be discolored by sapstains and molds. Sapstain caused by fungi (Fig. 37) is the blue, black, gray, or brown darkening of sapwood. The color is caused by large masses of dark fungal hyphae within the wood. Some sapstains may produce relatively bright colors such as red, purple, and yellow. Toughness is the only wood strength property seriously reduced by sapstains.

Molds (Fig. 38) discolor the surface of lumber. Molds are predominantly different shades of green, black, and occasionally orange or other light shades. This surface discoloration can generally be planed or even brushed off. On hardwoods, some shallow staining may result. As with stains, molds do not seriously affect strength properties other than toughness.

Molds and sapstains can be controlled by keeping wood dry (below about 20-percent moisture content). They are mentioned here because they indicate that high wood moisture contents that could permit serious wood decay are or have been present.

The wood decay described here should not be confused with pocket rot, which occurs in living trees and remains in the wood after it is manufactured into a product and put into service. Pocket rots do not continue to develop in dry lumber.

White pocket rots appear as spindle-shaped, pointed pockets or cavities parallel to the grain and separated by apparently sound wood. Within the pockets, the wood may be reduced to a white fibrous mass of cellulose, and in other cases, the pockets may be empty. This type of decay may be seen in Douglas Fir lumber cut from old growth trees. This "white speck" is accepted in limited amounts in some standard lumber grades because the large amount of wood between the pockets of decay is sound.

Brown pocket rots affect softwood species such as cedar and cypress. The pockets are elongated in the direction of the grain and are several times longer than they are wide. In the early stages of decay, the pockets are firm. In the advanced stages, the pockets are filled with a dark brown, carbonaceous, crumbly mass typical of brown cubical rot. The line of separation between the sound and decayed wood is sharp.

Figure 35—Discoloration from white rot on end grain, characterized by mottling and dark zone lines (arrows) bordering abnormally light-colored areas (top) and side grain of same board (bottom).

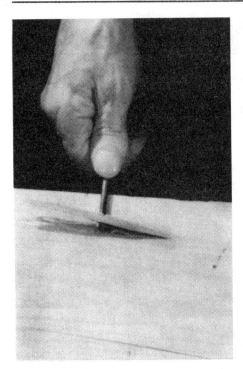

Figure 36—When wet wood is probed with a pick or comparable tool, it tends to lift out as a long sliver or break by splintering if sound (left). But if it is decayed even slightly, it tends to lift in short lengths and to break abruptly without splintering (right).

Figure 37—Sapstain and mold on sapwood of Southern Pine. Note that heartwood was not affected. Fungi have four requirements for growth: a food source (wood), oxygen, a favorable temperature, and moisture.

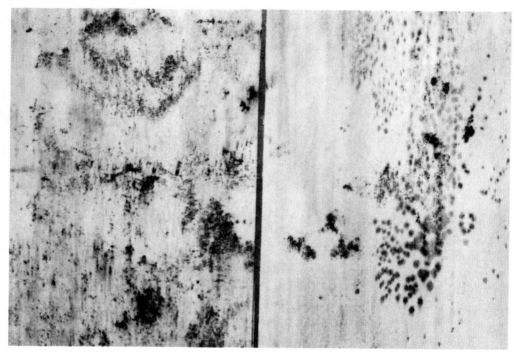

Figure 38—Surface mold.

As previously stated, decay occurs because excess moisture is present, either continuously or periodically. Excess moisture accumulates because of improper design or construction techniques, poorly made additions, or inadequate maintenance. Areas of wood structures that are exposed to or can accumulate moisture are more vulnerable to decay. Because the end grain of wood absorbs moisture faster than the side grain, decay often starts where two pieces are end-jointed together. In such joints, the wood becomes wet and does not dry out before decay fungi start growing.

Improper design or construction techniques can trap moisture and encourage decay. Placing untreated wood of nondurable species in exterior walls less than about 18 in. from the ground, where it will be exposed to splashing rain, can result in decay. Similarly, wood above crawl spaces should be not less than 18 in. above the soil. This problem commonly occurs at the bottom of side walls where sidewalks, porches, or decks join the side wall. Failure to cover the soil in the crawl space with a 6-millimeter plastic sheet will allow sufficient moisture vapor to pass from the soil to the wood above, where it may condense (Fig. 5). Wood wetted by this condensation can decay. Inadequate ventilation in the crawl space can aggravate this problem, particularly in corners with little air circulation.

Poorly ventilated attics, including those in which existing vents have been blocked by insulation, can also accumulate excess moisture. During cold weather, warm humid air moves from the living quarters into the attic. The moisture in the warmer air may condense, and if the space is not vented, decay in the roof decking and rafters can occur. Ice dams, formed when snow melts on the upper parts of a warm roof and flows to the eave and refreezes, can back water under the shingles, causing stains and possibly decay in ceilings and walls. Stains visible on the ceiling below the attic may indicate a moisture problem in the attic.

Clothes dryers, bathroom and shower areas, and kitchens should be vented to the outside. If they are vented to the crawl space or attic, excessive moisture is likely to accumulate. Because of the continual presence of moisture and the potential for faulty plumbing, wood located under and around bathrooms, showers, and laundry areas is often decayed. If these areas are located on the first floor, they can be easily checked from the basement or crawl space below. All seals and caulking around piping and vents should be in good repair.

Improperly installed flashing or lack of flashing around chimneys and at the junction of side walls and roof lines can cause decay. The edges of the roof line and fascia board are also vulnerable, particularly if the shingles do not extend about 1-1/2 in. beyond the edge.

Decay is common wherever wood contacts masonry material, such as where sill plates rest on a foundation. Masonry materials tend to retain water and transmit it to the wood. Using naturally durable wood or pressure-treated wood is suggested for these locations.

Water from wind-blown rain can accumulate on window sashes, doors, and frames, eventually causing decay, particularly at end-to-end joints. Wide roof overhangs tend to minimize this problem. Windows, sills, doors, and frames should be treated with a water-repellent preservative and painted to protect them from the weather.

Problems are often associated with changes in the design of homes and particularly where additions have been made. Be sure to examine these areas carefully for all the problems described previously.

Inadequate maintenance is another common source of excess moisture and subsequent decay. A leaky roof can cause decay in the roof decking and rafters. Make certain that the attic ventilation has not been blocked, particularly if insulation has been added. As discussed previously, stains on the ceiling below the attic may indicate a moisture problem in the attic.

Moisture may be present only during particular seasons. For example, wood in the crawl space may be excessively wet only in the winter or spring. Wood may therefore be moist enough to support decay for only short periods, and decay may occur at a relatively slow rate. Many older houses or parts of houses that were built prior to about 1930 were constructed with naturally durable heartwood. This wood will resist some decay even when wet. Less durable wood in the same type of construction is likely to decay more rapidly.

Insects

This section emphasizes evidence of insect activity because it is more easily observed than the insects themselves. While inspecting insect damage, determine how much structural damage has occurred, when the insect infestation occurred, and if the insects are still present. Pest control specialists can help to identify and control insects.

Conditions for Insects

Wood-destroying insects require three conditions to complete their life cycle and destroy wood: a source from which the infestation can spread, susceptible wood, and suitable temperature and humidity conditions. Insects are often selective in what they will attack and favor moist conditions that support fungal decay.

Insects that damage the wood in houses can be divided into two groups. The first group of insects, such as termites, carpenter ants, and powder-post beetles, generally infest wood some time after it is installed. These insects will continue to damage the wood until they are controlled. Powder-post beetles may be present in the wood before it is installed, then appear at a later date. The second group of insects infests standing or recently felled trees. Generally they are not active once wood is installed, but the damage remains obvious. Only the first group of insects will be discussed in this section. If damage is observed, probe the wood with a sharp tool to determine its extent.

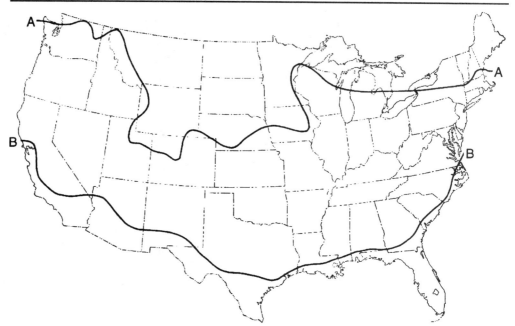

Figure 39—Range map for (A) subterranean and (B) drywood termites.

Termites

The two main categories of termites important in wood construction are subterranean termites, which require direct access to the ground or other water source, and drywood and dampwood termites, which do not require direct access to water. Subterranean termites are found through-out most of the United States, whereas drywood termites range in a narrow band from the warm, humid coastal areas of Virginia across the southern United States to central California (Fig. 39). Drywood termites are also found in Hawaii and the Caribbean islands.

Examine all areas of the house close to the ground for signs of subterranean termites. One of the most obvious signs is earthen tubes (Fig. 40) built on the surface of foundation walls to provide runways from the soil to the wood above. Termites may also enter through cracks or voids in foundations or concrete floors, through hollow-core concrete blocks, and through any earth-filled porches and flower planters.

Another sign of termites is the swarming of winged adults early in the spring or fall. Termites resemble ants, but termites have much longer wings and do not have the thin waist of an ant (Fig. 41).

Termite-damaged wood shows several characteristics when broken open (Fig. 42). First, termites tend to eat the soft earlywood, leaving the hard latewood. These eating habits form galleries that follow the grain of the wood. A thin outer shell is left around the entire piece of wood. Some cavities may contain substantial quantities of soil mixed with chewed wood.

The only other visible sign of termite infestation is the presence of dark spots or blister-like areas on flooring, trim, or framing. These areas are easily penetrated with a knife or screw-driver. In cases of severe damage, a wood member might be partially collapsed at bearing points or might have otherwise failed. Sometimes internal damage can be determined by probing the wood with a sharp instrument or by pounding with a hammer to detect hollow areas.

Many older houses, particularly in the South, are equipped with termite guards. Termite guards are usually galvanized metal sheets placed between the wood sills and the foundation or footings. Termites can still build tubes around them and infest the house, particularly if the shields were not properly installed.

70

Figure 40—Subterranean termite shelter tubes on interior foundation wall of crawl space.

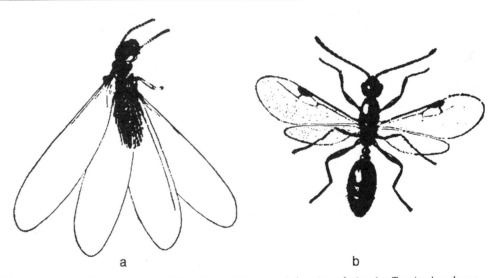

a   b

Figure 41—Major differences between (a) termites and (b) ants include waist and wing size. Termites have larger wings and nonrestricted waists.

Subterranean termites are usually controlled with insecticides. Treating the soil under and around the foundation with an approved insecticide is probably the most effective way to protect buildings. Good design and construction practices should also be followed. Untreated wood on wall surfaces should be at least 8 in. from the soil line and definitely not in contact with the soil. Wood in crawl spaces should be at least 18 in. above the soil. Earth-filled porches, flower planters, and such, should not be attached directly to the house. Wood debris should not be buried under or near the house. Initial damage from a termite infestation is slow, but as the colony develops, it can become severe.

Figure 42—Termite-damaged wood showing the insects' preference for soft earlywood and accumulation of soil and fecal material in galleries.

Drywood termites tend to work just under the surface of the wood, leaving a very thin veneer-like layer. Wood damaged by drywood termites has broad pockets or chambers connected by tunnels that cut across the grain without regard for earlywood or latewood (Fig. 43). The galleries are perfectly smooth and have no surface deposits. Some fecal pellets may be stored in portions of the galleries; the galleries are closed off by partitions made of fecal pellets stuck together with a secretion.

Piles of fecal pellets are usually the first sign of a drywood termite infestation. The pellets are hard, elongate, and less than 1/25 in. long, with rounded ends and six flattened or concavely depressed sides. They are light gray to very dark brown in color. The pellets are eliminated from the galleries in the wood through round kick holes. Unused kick holes are closed with a secretion and pellets. Probing wood with a sharp instrument or pounding the surface may reveal hidden damage.

Methods for preventing drywood termite attack are not as practical nor as economical as those to protect against subterranean termites. However, control measures should be taken where drywood termites are considered to be a serious problem (Pacific area, southern coastal California, southern Florida, and Caribbean area).

Dampwood termites build their colonies in damp and sometimes decaying wood. Soil contact is not required if the wood is damp. Although these termites are found in the southern tips of Florida, the Caribbean, Nevada, Idaho, and Montana, their largest economic impact is on the Pacific coast.

Carpenter Ants

Carpenter ants are typical of all ants. They have a very narrow waist (unlike termites) and wings of two sizes, the front ones much larger than the back ones (unlike termites, which have equal-sized wings). Ants and termites are compared in Figure 41. The adult ants found nesting in houses are predominantly black; however, some may be partially reddish-brown to yellowish.

Figure 43—Drywood termite damage showing round "kick holes" and very thin veneer-like layer left after excavating without regard for earlywood or latewood.

Carpenter ants burrow into the wood to make nests, but do not use the wood for food. Most species prefer to nest in moist wood that has begun to decay. The galleries extend both along the grain of the wood and across the annual rings. The ants tend to remove the softer earlywood first. They penetrate the harder-grained latewood at frequent intervals, so that, unlike galleries made by subterranean termites, there is complete access (across annual rings) between the galleries (Fig. 44). Once a nest is established, it can be extended into the sound wood. The surfaces of ant galleries are smooth and perfectly clean; this contrasts with the galleries of subterranean termites in which the walls are rough because of the coating of fecal materials.

Other indications of carpenter ant infestation are piles or scattered bits of very fibrous and sawdust-like frass that the ants have removed from the wood. The frass is expelled from cracks and crevices, or from slit-like openings, called windows, made by the ants. These windows are the only external evidence of attack by carpenter ants. The frass is quite often found in basements, dark closets, attics, under porches, and other out-of-the-way places. Faint rustling and even gnawing sounds can be heard in the wood or cavity when the ants are active.

Carpenter ant nests should be treated with an approved residual contact insecticide registered for treating carpenter ants. The infested wood can be bored, the pesticide injected, and the

Figure 44—Carpenter ant damage to Douglas-fir showing preference for earlywood and removal of some latewood to allow access between galleries.

holes plugged. The approaches and areas surrounding the nest should also be treated. Treating the areas where the ants are visible is seldom effective because many ants never leave the nest. Poison baits may also be available. Good building practices reduce the risk of attack by carpenter ants.

Powder-Post Beetles

The term powder post describes a group of beetles that convert the inner portion of infested wood to a powdery or pelleted mass. The thin outer shell on the surface of infested wood often shows numerous small holes through which the beetles have exited. Damage tends to be heaviest in warm, humid climates or in houses with moisture problems. Powder-post beetle damage in older homes is not uncommon. However, the infestations are not always active. Current activity is indicated by fresh dust streaming from the holes or dust accumulating in piles below the holes. Older boring dust may be yellow or partially caked on the surface.

If beetle damage is active, increase ventilation to reduce the moisture content of the wood and create an environment that is less favorable to beetle development. If the infestation is extensive and needs to be controlled, the area can be fumigated.

Bostrychids and Flat-Headed Beetles

These and some other insects can infect wood that has just been cut or is seasoning. They are generally found where bark is left on the edges of the lumber. The entry holes, with 3/32- to 9/32-in. diameters, are larger than the holes created by powder-post beetles and are filled with meal-like, tightly packed frass. These insects normally do not reinfest the wood. One generation usually takes 1 year, but can take 5 years, to mature. The insects generally do not damage wood substantially, but their emergence can concern homeowners.

Rodents

If proper precautions are taken, rodents are usually not a serious problem in homes. However, when the structure has been neglected and if an abundant food source is available, rodent populations can grow rapidly. Serious damage and health hazards can result.

Rat and mouse damage is visible through gnawings, droppings, urine stains, tracks, burrows, and smudge or rub marks. Smudge or rub marks are found along frequently traveled routes and are the result of oil and dirt that is rubbed off the rodent's fur. Rats cause damage by gnawing through structural members, doors, windows, floors, and other areas, and by burrowing under concrete, causing settling and cracking. They also gnaw on electrical wiring and even water pipes. Both rats and mice can destroy insulation. Rodents can also transmit disease to humans.

Rats can enter a building through any opening that is 1/2 in. wide or more; mice need an opening of just 1/4 in. Rodents' ability to penetrate through small openings, their agility in climbing and jumping, and their ability to burrow and gnaw through various materials, including lead, aluminum sheeting, wood, rubber, vinyl, and even concrete block, makes it difficult to bar them from an older structure.

The county agricultural extension service can suggest ways to control rodents and other animals. Rodenticides and trapping are possible control methods. Pest control operators can also assist.

In addition, good sanitation practices reduce the sources of food, water, and shelter. Foodstuffs should be stored in rodent-proof containers. Garbage should be disposed of properly, and weeds and debris should be removed from around the structure.

Rehabilitation

Introduction

This portion of the handbook provides information and guidance for the actual work of rehabilitating an older house.

"Hazard Control" in the "Rehabilitation" section discusses measures that can control the four health and safety hazards that are described in the "Condition Assessment" section: radon, lead, asbestos, and fire.

The "Rehabilitation" section also includes sections on thermal protection and moisture control.

The ensuing sections provide information on the external and internal rehabilitation of older dwellings, feature by feature and area by area. Beginning with site improvements and major structural features, there is information on siding, exterior finishing, interior structural changes, interior finishing, vertical and horizontal expansion, kitchen improvements, utilities, fireplaces and chimneys, garages and carports, and deck and porch additions.

As with the material on condition assessment, the information in this section is designed to be useful to professional remodelers and people in related trades, persons who are rehabilitating an older home on a part-time or spare-time basis, and students interested in architecture and building. Although many of the procedures involved in rehabilitating an older house lie beyond the skills and experience of the average home dweller, this text is designed to be understandable to the layperson who would like to know what is being done in the rehabilitation of the home and how it is being done.

Hazard Control

Radon

Introductory information on radon and on methods for measuring radon levels in homes is provided in "Radon" under "Health and Safety Assessment" in the "Condition Assessment" section.

In its publication *A Citizens Guide to Radon,* the EPA uses a radon level of 4 pCi/L (0.02 WL) as a benchmark below which radon mitigation action is probably not necessary. For the meaning of these terms, see "Radon Detectors" under "Health and Safety Assessment" in the "Condition Assessment" section.

The EPA recommendations for dealing with elevated radon levels are as follows:

- Readings of about 4 pCi/L (0.02 WL) or lower.

    Exposures in this range are considered average or slightly above average for residential structures. Although people who are exposed to this range are not totally free from risk of lung cancer, it may be difficult or impossible to reduce levels below these readings.

- Readings of about 4 to 20 pCi/L (0.02 to about 0.1 WL).

    Exposures in this range are considered above average for residential structures, and the EPA recommends that levels be brought down to 4 pCi/L (0.02 WL) or below. Such action should be taken within a few years of occupancy, or sooner if levels are at the upper end of this range.

- Readings of about 20 to 200 pCi/L (0.1 to 1.0 WL).

  Exposures in this range are considered greatly above average for residential structures. Action should be undertaken within a few months to reduce levels as far below 20 pCi/L (0.1 WL) as possible.

- Readings of about 200 pCi/L or higher (1.0 WL or higher).

  Such exposures are among the highest observed in homes. Action should be taken within a few weeks to reduce levels as low as possible. If immediate action is not possible, the homeowner or occupant should consult with State or local health or radiation protection officials to determine if temporary relocation is appropriate until the levels can be reduced.

Radon Entry

Radon levels in soils vary from place to place and can be different only a few feet apart. Older houses of similar appearance and construction can have varying susceptibility to radon entry. Therefore, radon mitigation plans must be individually developed. Some methods are simple, some are complex, and some are much more expensive than others.

Figure 45 shows various ways that radon enters a home. Radon can be emitted from materials such as granite and shale used in interior applications such as a fireplace (item J in Fig. 45). Radon can also be present in private wells and can be released by water flow in the home (item K in Fig. 45). However, these are unusual circumstances, and most radon that enters homes comes from the soil.

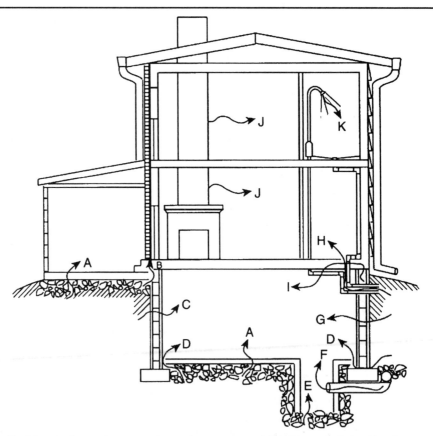

Figure 45—Major radon entry routes: (A) cracks in concrete slabs, (B) spaces behind brick veneer walls that rest on uncapped hollow-block foundation, (C) pores and cracks in concrete blocks, (D) floor-wall joints, (E) exposed soil, as in a sump, (F) weeping (drain) tile, if drained to open sump, (G) mortar joints, (H) loose-fitting pipe penetrations, (I) open tops of block walls, (J) building materials such as some rock, and (K) water (from some wells).

A number of mitigation techniques require the services of a professional contractor who has skills and experience in the field. Many states provide lists of such contractors, and some states have certification programs. The homeowner should secure a written proposal from one or more contractors. They should request and check references. The contractors' business reputations can be checked with the Better Business Bureau or Chamber of Commerce. A second opinion from another contractor, or counsel from a state or local radiological health official, can be helpful in deciding if a proposal is reasonable.

The EPA publication *Radon Reduction Methods: A Homeowner's Guide* lists various mitigation techniques, including the following:

- Natural ventilation and forced ventilation.

  These techniques involve opening basement windows and either allowing natural airflow or providing forced-air entry into the basement space by placing a fan in one window. The techniques are highly effective, but their use is limited by heat and cooling loss and security problems.

- Heat recovery ventilation.

  A heat recovery ventilator, also called an air-to-air heat exchanger, replaces indoor air with outdoor air, thereby venting radon to the outside.

- Covering exposed earth.

  This includes such measures as covering exposed earth under a crawl space with polyethylene or some other barrier, pouring a concrete floor over an earthen basement floor, and providing a tight cover for sumps.

- Sealing cracks and openings.

  These include openings around utility pipes, joints between basement floors and walls, including perimeter (French) drains, and holes in the top row of concrete blocks.

- Draintile suction.

  Water is drained away from the foundation of some houses by perforated pipes called draintiles. If this pipe system forms a partial or continuous loop around the house, it can be joined to a fan and used to pull radon from the surrounding soil and vent it away from the house.

- Sub-slab suction.

  The lowest floor of most homes consists of a concrete slab. If this slab has been poured over aggregate or rests on highly permeable soil, one or more pipes can be sunk beneath the slab into the aggregate or soil, joined to an exhaust fan, and vented outside the house. Similar methods can be used in crawl spaces without slabs.

- Block-wall ventilation.

  If basement or crawl space walls are constructed of concrete masonry blocks, the hollow cores in the block form an interior network that can be either pressurized by a fan to prevent radon entry or exhausted by a fan to remove radon.

- Prevention of house depressurization.

In many homes, air pressure inside the home is lower than that outside. This has the effect of drawing radon into the house. Radon entry can be reduced by keeping the house basement or heated crawl space under positive pressure with a fan.

Table 5 shows various radon mitigation methods. More complete information on each technique, along with a discussion of its efficiency and likely range of cost, is contained in *Radon Reduction Methods*.

Table 5—Comparison of features for various methods of radon mitigation [a]

Method	Installation cost	Operating cost	Maximum possible reduction[b]	Comment

Natural ventilation:
  Basement or
  lowest floor
	Minimal	High to
		very high
			Up to 90+%	Useful step to reduce high
				radon levels immediately.

Crawl space	Minimal	Moderate	Up to 90+%	

Forced ventilation:
  Basement or
  lowest floor
	Low to moderate	Very high	Up to 90+%	More controlled than natural
				ventilation.

Crawl space	Low to moderate	Moderate	Up to 90+%	
Heat recovery ventilation: Ducted	Moderate to high	Low to moderate	50%–75%	Air intake and exhaust must be equal. Also, expect lower radon reductions for houses with moderate to high air-exchange rates.
Wall-mounted	Low to moderate	Low to moderate	No data available	
Covering exposed earth	Moderate to high	Low	Site-specific	Required to make most other methods work.
Sealing cracks and openings	Minimal to high	Nominal	Site-specific	Required to make most other methods work.
Draintile suction	Minimal to high	Low	Up to 90+%	Works best when drain is a continuous, unblocked loop.
Sub-slab suction	High	Low	Up to 90+%	Works best with good aggregate or highly permeable soil under slab.
Block-wall ventilation	High	Low	Up to 90+%	Applies to block-wall basements. Sub-slab suction may be needed to supplement.
Prevention of house depressurization	Low to moderate	Low	Site-specific	May be required to make other methods work. May see seasonal impact.
House pressurization	Moderate to high	Moderate	Up to 90+% (limited data)	Most cost-effective when basement is tightly sealed.

[a]From EPA, *Radon Reduction Methods—A Homeowner's Guide,* September 1987.

[b]These represent generally the best reductions a single method can accomplish. Reductions may be higher or lower, depending on the unique characteristics of the house. Especially with high initial radon levels, several methods may have to be combined to achieve acceptable results.

| Lead | Introductory information on lead contamination in the home and on problems of lead-based paint abatement appears in "Lead" under "Health and Safety Assessment" in the "Condition Assessment" section. |

The Agency for Toxic Substances and Disease Registry estimates that there are about 42 million homes in the United States constructed before 1980 that have lead-based paint on their interior or exterior. As more and more of these homes are rehabilitated, the problem of lead-based paint abatement has become the subject of increased public and official concern.

Lead-based paint on the exterior of homes chalks and powders because of moisture and ultraviolet light. The extremely fine lead dust can accumulate in the soil near the house and can ultimately enter the house. Poor quality lead-based paint used on interior surfaces of the home can also produce dust. Lead dust can be generated when coatings on surfaces are broken because of aging or rehabilitation. Finally, the process of abatement itself can generate lead dust. This is particularly true when unacceptable methods and work practices are used. A poorly performed abatement can be worse than no abatement at all. The micrometer-sized lead dust particles, like asbestos fibers that are of similar size, can remain airborne for substantial periods, cannot be completely removed by standard cleaning methods, and can re-enter the air when disturbed. The particles are easily ingested by the body and can cause elevated levels of lead in the blood, which can be dangerous to human health as described in "Lead" under "Health and Safety Assessment" in the "Condition Assessment" section.

Research on Lead-Based Paint Abatement

As of this writing, there are no standardized technical guidelines for lead-based paint abatement, and the subject represents a frontier of research and development.

The Housing and Community Development Act of 1987 (PL 100–242) directed HUD to conduct a demonstration program to determine what abatement techniques to use in HUD-owned housing. Under the program, HUD will

- Secure information on the reliability and accuracy of available technology to test lead-based paint and develop recommendations for the best detection technique.

- Develop detailed information on the cost and safety considerations associated with various methods of abatement.

- Prepare an estimate of the regional distribution of housing in the United States that contains lead-based paint.

At the conclusion of the program, HUD must submit a report and a plan for the prompt and cost-effective inspection and abatement of lead-based paint in privately owned single family and multifamily housing.

When it is released, this report will provide important guidance for lead-based paint abatement in the rehabilitation of older homes. The local HUD office can be contacted about the status of this program.

Regulatory Activity

State and local environmental health agencies are beginning to enter the field of lead-based paint abatement and can be expected to play an increasing role. Two states, Maryland and Massachusetts, promulgated regulations for lead-based paint abatement in 1988. Other states and jurisdictions are likely to mandate allowable abatement procedures and testing methods.

Advice and Assistance	Rehabilitation activities that will require disturbing, removing, or demolishing portions of the structure that are coated with lead-based paint pose serious problems on which the home dweller should seek information and advice. Professional assistance may be desirable.

State and local environmental health offices can provide information on rules and regulations for testing, abatement, post-abatement cleanup, and disposal of debris. With regard to the latter, remember that debris coated with lead-based paint may be regarded as hazardous waste.

Local building inspectors and local home building associations can provide the names of contractors who have knowledge of and experience with lead-based paint abatement.

Testing for Lead-Based Paint

The Lead-Based Paint Poisoning Prevention Act (LPPPA), initially passed by Congress in 1971 and amended several times in succeeding years, requires HUD to abate all interior and exterior surfaces in public and Native American Indian housing with a lead content of 1.0 mg/cm^3 or higher. The Maryland and Massachusetts statutes require abatement in public housing under their jurisdictions when the level of lead in paint is 0.5 percent or higher. The two standards are not directly comparable; in most instances, the 0.5-percent standard is more stringent.

A portable X-ray fluorescence (XRF) analyzer is commonly used to determine lead levels in paint. Because this device can give very inaccurate results if used by a person not experienced in its operation, the analysis should be done by a qualified professional.

Chemical spot testing, using a solution of 6- to 8-percent sodium sulfide, is sometimes used to screen painted surfaces for the presence of lead. However, this approach is not recommended by the National Institute of Building Sciences Lead-Based Paint Project Committee because it is not always reliable.

Unacceptable Abatement Methods

Traditionally, lead-based paint removal methods generated high levels of particulate dust. These methods were pursued with little effort to contain the dust and were characterized by inadequate worker protection. In many instances, the treated surfaces to which the particles adhered were not repainted. Disposal of debris was haphazard. Various Federal and State regulations have prohibited the following methods, so they should *not* be employed because they generate lead dust:

- *Use of gas-fired open-flame torches*

- *Grinding or sanding without attached high-efficiency particulate air (HEPA) vacuum filtration apparatus*

- *Uncontained water blasting*

- *Open abrasive blasting*

Dry-scraping is prohibited by Maryland abatement regulations, but is currently allowed by HUD. To minimize dust generation, dry-scraping should be accompanied by misting and should be used for limited areas.

Acceptable Abatement Strategies

Four acceptable strategies for lead-based paint abatement are

- replacement of painted components,

- sealing off of painted components or areas,

82

- paint removal on-site, and

- paint removal off-site.

Table 6 summarizes information relating to each of the four strategies.

Preparatory Procedures

The following preparatory procedures should be performed before carrying out interior lead-based paint abatement or rehabilitation activities that will disturb lead-based paint:

- Remove furniture, draperies, and wall-to-wall carpeting from the area where abatement or rehabilitation will take place.

- Seal off the work area from other areas of the house with 6-mil polyethylene or equivalent barriers attached to framing or by other suitable means.

- Cover all floors in the work area with 6-mil polyethylene or equivalent sheeting.

- Shut down forced-air heating and air-conditioning systems, and seal all air intake and exhaust points of these systems.

On-Site Removal Methods

The methods described for on-site removal should be restricted to limited surface areas:

- Heat-based removal methods.

  High levels of airborne lead can be produced by heat guns, making the use of respirators essential. At the temperatures at which most heat guns operate, some lead is likely to be volatilized. Substantial lead is volatilized and lead fumes are released at about 700°F. Heat guns capable of reaching or exceeding 700°F should not be operated in that range.

- Chemical removal methods.

  Chemical substances used for paint removal are hazardous and should be used with great care. Solvent-based chemical strippers are flammable and require ventilation. They may contain methylene chloride, a central nervous system depressant that can cause kidney and liver damage at high concentrations and is a probable carcinogen. Use supplied-air respirators when working with strippers containing this substance. When solvent-based strippers that do not contain methylene chloride are used, add organic vapor filters to respirators.

  Chemical removal may require multiple applications, depending on the number of layers of paint. Caustic and solvent-based chemicals should not be allowed to dry on the lead-painted surface. If the chemical dries, paint removal will not be satisfactory, and the potential for creating lead dust will be increased.

  Caustic chemical strippers have a very high pH (alkaline content), which can cause severe skin and eye injuries.

- Mechanical removal methods.

  Scraping should be performed with misting. Sanding without a HEPA-filtered vacuum should not be the finishing method after scraping or any other method of abatement. HEPA sanders are recommended only for limited surface areas; they are most appropriate for flat surfaces such as jambs and stair risers.

Table 6—Comparison of lead-based paint abatement strategies [a]

Abatement strategy	Advantage	Disadvantage	Appropriate application	Inappropriate application	Additional information
Replacement	Permanent solution Allows upgrade Can be integrated with modernization No lead residue left behind on surfaces Low risk of failing to meet clearance standards	Replacement component may be lesser quality than original Replaced components may be high volume and considered hazardous waste Certain installation requires skilled labor	Many interior or exterior components Deteriorated component Highly recommended for windows, doors, and easily removed building components	Restoration projects When historic trust requirements apply Most walls, ceilings, and floors	Nonstandard replacement components may need to be ordered in advance Demolition may damage surfaces May result in increased energy efficiency, for example, if windows are replaced
Encapsulation	Minimal dust if surface preparation is minimal May be faster than some other methods	May not provide long-term protection Requires routine inspection May require routine maintenance Quality installation critical for durability	Exterior trim, walls, floors Interior floors, walls, ceilings, pipes Balustrades	When encapsulant is not appropriate for substrate condition	Encapsulant must be durable and seams must be sealed to prevent escape of lead dust Safe, effective, and aesthetic encapsulants for interior trim components need to be developed and tested Repainting leaded surfaces and using contact paper and paper wall coverings should not be considered for abatement
On-site paint removal	Low level of skill required Allows restoration	Much dust generated Lead residue may remain on substrate and may be difficult to remove Potential difficulty in meeting clearance standards and in protecting workers Stripping agents are hazardous and require more precautions	Limited surface areas When replacement, encapsulation, and off-site removal are impractical	Large surface areas	Unacceptable paint removal methods: – gas-fired open-flame burning – grinding or sanding without HEPA[b] filtration – dry-scraping without misting – uncontained water blasting – open abrasive blasting Chemical removers work best on metal substrates Check with chemical manufacturer regarding recommendations for use on various types of wood and metal substrates
Off-site paint removal	Allows restoration Better finished product generally than with on-site paint stripping	Lead residue may remain on substrate, which may be difficult to remove Damage may occur during removal and reinstallation Swelling of wood, glass breakage, and loss of glues and fillers may occur Hardware left on components may be damaged	Restoration projects, especially doors, mantels, easily removed trim Metal railings	– – –	Check with stripping company for timing of work and procedures for neutralizing and washing components Check with stripping contractor regarding recommendations for metal substrates

[a]Source: National Institute of Building Sciences, 1989.
[b]HEPA is high-efficiency particulate air vacuum filtration apparatus.

Of the four strategies listed in Table 6, only encapsulation is recommended for abatement of large interior surfaces such as walls, floors, and ceilings. Such encapsulation involves covering and sealing of the surfaces with such materials as gypsum board, wood, plywood, or plywood paneling. Seek guidance from abatement professionals where abatement may be desirable, but covering walls, floors, or ceilings is not desirable or practical.

Asbestos

Health hazards associated with asbestos are described in "Asbestos" under "Health and Safety Assessment" in the "Condition Assessment" section. Building materials containing asbestos were used in the construction of many older homes. Commonly used materials of this type include sprayed-on or troweled-on wall and ceiling coatings, wall and ceiling insulation, preformed pipe wrap, and vinyl asbestos floor tile. Table 7 is a detailed list of asbestos-containing materials (ACMs) used in home construction.

A reference on dealing with asbestos in the context of home remodeling and rehabilitation is the book *Asbestos Handbook for Remodeling*, published jointly by the National Association of Home Builders (NAHB) and the Association of the Wall and Ceiling Industries.

As stated in "Asbestos" under "Health and Safety Assessment" in the "Condition Assessment" section, there are two types of ACMs, *nonfriable* and *friable*. The asbestos fibers in nonfriable ACMs are embedded in a matrix of other material and will not be released under ordinary conditions of use. Friable ACMs are soft and can easily be crumbled, releasing fibers into the atmosphere. This type of ACM poses the greatest health hazards and is the principal focus of Federal, State, and local regulations.

Leading asbestos abatement firms frequently encounter dangerous conditions that have been created in the process of home rehabilitation. With only a few exceptions that are described in this section, *home dwellers and remodelers who are not trained and certified in the field of asbestos abatement should not attempt to work with asbestos.*

When asbestos abatement is required in home rehabilitation, the services of a trained and accredited abatement contractor should be secured. Asbestos abatement has become a highly technical business that is closely regulated at the federal, state, and local levels. Untrained and unaccredited persons who attempt to work on asbestos abatement can create health hazards for themselves and others. They also expose themselves to the risk of serious legal penalties.

Asbestos Regulations

Major regulatory programs and activities for asbestos handling and abatement, arranged by organization, are as follows:

1. EPA

a. Prohibition of use

In 1973, EPA prohibited spraying ACMs for insulation, fireproofing, and soundproofing. In 1975, the Agency prohibited the use of asbestos for pipe covering if the material could be easily crumble after it dried. In June 1978, EPA banned the use of all other friable ACMs.

b. Required notification

The EPA administers the National Emission Standards for Hazardous Air Pollutants (NESHAPS), which were established under the Clean Air Act. The NESHAPS rules state that, when a building is demolished or if more than 260 linear feet of asbestos pipe insulation or

Table 7—Location, composition, and dates of use of asbestos-containing building products [a]

Product	Location	Percent asbestos	Dates of use	Binder	Friable/nonfriable	How fibers can be released
Roofing felts	Flat, built-up roofs	0–15	1910–present	Asphalt	Nonfriable	Replacing, repairing, demolishing
Roof felt shingles	Roofs	1	1971–1974	Asphalt	Friable	Replacing, demolishing
Roofing shingles	Roofs	23–32	?–present	Portland cement	Nonfriable	Replacing, repairing, demolishing
Roofing tiles	Roofs	20–30	1930–present	Portland cement	Nonfriable	Replacing, repairing, demolishing
Siding shingles	Siding	12–14	?–present	Portland cement	Nonfriable	Replacing, repairing, demolishing
Clapboards	Siding	12–15	1944–1945	Portland cement	Nonfriable	Replacing, repairing, demolishing
Sprayed coating	Ceilings, walls, and steelwork	1–95	1935–1978	Portland cement, sodium silicate, organic binders	Friable	Water damage, deterioration, impact
Troweled coating	Ceilings, walls	1–95	1935–1978	Portland cement, sodium silicates	Friable	Water damage, deterioration, impact
Asbestos-cement sheet	Near heat sources such as fireplaces, boilers	20–50	1930–present	Portland cement	Nonfriable	Cutting, sanding, scraping
Spackle	Walls, ceilings	3–5	1930–1978	Starch, casein, synthetic resins	Friable	Cutting, sanding, scraping
Joint compound	Walls, ceilings	3–5	1945–1977	Asphalt	Friable	Cutting, sanding, scraping
Textured paints	Walls, ceilings	4–15	?–1978	Asphalt	Friable	Cutting, sanding, scraping
Millboard, rollboard	Walls, commercial buildings	80–85	1925–?	Starch, lime, clay	Friable	Cutting, demolishing
Vinyl wallpaper	Walls	6–8	?		Nonfriable	Removing, sanding, dry-scraping, cutting
Insulation board	Walls	30	?	Silicates	Friable	Removing, sanding, dry-scraping, cutting
Vinyl-asbestos tile	Floors	21	1950–1980?	Poly(vinyl)chloride	Nonfriable	Removing, sanding, dry-scraping, cutting
Asphalt-asbestos tile	Floors	26–33	1920–1980?	Asphalt	Nonfriable	Removing, sanding, dry-scraping, cutting
Resilient sheet flooring	Floors	30	1950–1980?	Dry oils	Nonfriable	Removing, sanding, dry-scraping, cutting
Mastic adhesives	Sheet and tile backing	5–25	1945–1980?	Asphalt	Friable	Removing, sanding, dry-scraping, cutting
Cement pipe and fittings	Water and sewer mains	20–?	1935–present	Portland cement	Nonfriable	Demolishing, cutting, removing
Block insulation	Boilers	6–15	1890–1978	Magnesium carbonate, calcium silicate		Friable damage, cutting, deterioration
Preformed pipe wrap	Pipes	50	1926–1975	Magnesium carbonate, calcium silicate		Friable damage, cutting, deterioration
Corrugated asbestos paper						
High temperature	Pipes	90	1935–1980?	Sodium silicate, starch	Friable	Damage, cutting, deterioration
Moderate temperature	Pipes	35–70	1910–1980?	Sodium silicate, starch	Friable	Damage, cutting, deterioration
Paper tape	Furnaces, steam valves, flanges, electrical wiring	80	1901–1980?	Polymers, starches, silicates	Friable	Tearing, deterioration
Putty (mudding)	Plumbing joints	20–100	1900–1973	Clay	Friable	Water damage, cutting, deterioration

[a]Source: U.S. Environmental Protection Agency.

160 ft^2 of asbestos surfacing material are removed during remodeling or renovation, the owner must give 10 days advance notice to the EPA regional office and/or state environmental office. If the amount of ACMs to be removed is less than these amounts, 20 days notice must be given. The notification must include

- starting and completion dates of the work;

- description of the planned removal methods;

- the name, address, and location of the disposal site; and

- any additional information and reports required by state and local regulations.

c. Training and accreditation

In 1986, Congress passed an amendment to the Toxic Substances Control Act, called the Asbestos Hazardous Emergency Response Act (AHERA), that requires asbestos surveys, reports, and management plans for control and abatement of ACMs in schools.

One section of this act places responsibility on the states for setting up training and accreditation programs for inspectors, management planners, project designers, contractors, and workers in the field of asbestos handling and abatement. Accreditation certificates are granted to persons who take the courses and pass written examinations. The courses are open to all persons, including do-it-yourself homeowners involved in construction or rehabilitation of buildings and homeowners who may wish to have the relevant information and to be accredited.

The EPA has funded training programs at eight universities, and has approved about 250 courses and 140 course providers throughout the country. In addition to the EPA-funded training programs, EPA-approved providers of training in asbestos abatement include schools, laboratories, consulting firms, associations, private vendors, and building trade unions. Names of approved providers can be obtained from EPA and from state and local environmental offices.

2. Occupational Safety and Health Administration

In July 1986, the Occupational Safety and Health Administration (OSHA) issued regulations requiring employers, including home rehabilitation contractors whose employees encounter asbestos, to undertake specified procedures to protect their workers and the public. The regulations are extensive and complex. The compliance burden falls on the employer. The OSHA's requirements are summarized in the *Asbestos Handbook for Remodeling*.

a. State and local regulations

Federal authorities regard many of their regulations as representing minimal requirements, and federal policy encourages supplemental state and local regulation. A number of states and some local jurisdictions have enacted asbestos regulations with which homeowners and rehabilitation workers must comply. Information on these requirements can be obtained from state and local environmental offices.

There are three strategies for dealing with asbestos in homes: encapsulation, enclosure, and removal.

- Encapsulation

  Encapsulation involves applying chemical sealants to ACMs to bind the asbestos fibers together and seal them in. There are two types of encapsulants—penetrating and bridging—that may be used singly or, at times, together. Penetrating encapsulants are adhesives that are absorbed by the ACMs, binding the fibers firmly inside the material and leaving a permeable surface coating. Bridging encapsulants, which can be applied directly over a penetrant, dry on the surface to form a tough, airtight, waterproof seal.

- Enclosure

  Enclosure involves constructing airtight walls and ceilings around ACMs, then covering the exterior of the enclosure with impact-resistant materials. The barrier can be constructed of various materials, including gypsum board, tongue-and-groove plywood, concrete, masonry, or metal.

- Removal

  Removal has the advantage that, if it is done properly, the premises are henceforth free of ACMs. However, the procedure may involve the greatest immediate cost. If not done by knowledgeable professionals, removal poses high risks of increasing the airborne asbestos count on the premises. Even when the waste is taken to an approved or regulated landfill, the owner retains responsibility for it perpetually.

In some instances, it is possible to conduct certain types of small-scale, short-duration rehabilitation work involving asbestos without professional assistance. As a guide to defining small-scale work, persons working on home rehabilitation can use the criterion of 3 ft^2 of surface or 3 linear feet of material, such as asbestos pipe, which appears in one of EPA's regulations (40 CFR 763, subpart G).

In conducting work that falls within these guidelines, EPA recommends thoroughly wetting the ACMs to be disturbed. Vacuum the area with HEPA (high-efficiency particulate air) vacuuming equipment and place wet waste in marked, doubled 6-mil polyethylene bags as quickly as possible. If the ACMs are friable, the bags of waste should be disposed of in an EPA or state-approved dump or landfill. Obtain a dump receipt and keep it permanently. If the waste is nonfriable and not extensively broken up, it may be disposed of in any landfill, although some state and local laws may prohibit this.

For certain types of small-scale, short-duration rehabilitation and maintenance activities, OSHA permits an optional exemption from its full requirements. Such activities, OSHA states, include but are not limited to the following:

- Removal of asbestos-containing insulation on pipes

- Removal of small quantities of ACMs on beams and above ceilings

- Replacement of an asbestos-containing gasket on a valve

- Installation or removal of what OSHA describes as "a small section of drywall"

- Installation of electrical conduits through or proximate to ACM.

According to OSHA, before workers perform such operations, all movable objects should be removed from the area to avoid contamination. Objects that cannot be moved must be covered completely with 6-mil polyethylene plastic sheeting. Objects that may already have been contaminated should be thoroughly cleaned with an HEPA vacuum or wet-wiped before they are removed from the area, or they should be completely encased in plastic.

The rule states that, whenever feasible, wet methods should be used in working with ACMs; that is, apply water or another wetting agent with an airless sprayer. "Only in cases where asbestos work must be performed on live electrical equipment, on live steam lines, or in other areas where water will seriously damage materials or equipment, may dry removal be performed [with prior approval]," according to OSHA.

The rule contains detailed information on using glove bags for small-scale, short-duration removal of ACMs from pipes and similar structures.

With regard to removing old vinyl asbestos floor tiles and flooring, OSHA states, "There appears to be virtually no possibility that the excursion limit would be exceeded if the recommendations of the Resilient Floor Covering Institute were followed." The excursion limit, defined as one asbestos fiber per cubic centimeter of air, averaged over 30 min is the threshold for triggering OSHA regulations. The Institute's recommended procedures are contained in the publication *Recommended Work Procedures for Resilient Floor Coverings*.

Fire Safety

Information on assessing older dwellings for fire safety appears in "Fire Safety" under "Health and Safety Assessment" in the "Condition Assessment" section. Ways to increase fire safety are described in this section.

House Design and Layout

Older persons are at much higher risk of dying in residential fires than is the general population. Pay particular attention to the problem of emergency egress for older persons and handicapped persons whose mobility may be limited.

Establishing a bedroom on the first floor for an older resident or handicapped resident can be desirable in a rehabilitation plan. Such a room can often be created from an existing dining room or study with little physical alteration of the space and with little or no loss of living convenience on the first floor. A door to the room can be added if one is not present.

Because the converted room was not originally a bedroom, it is important that the standards of the Council of American Building Officials (CABO) One- and Two-Family Dwelling Code for emergency egress openings from sleeping rooms be followed, and appropriate changes made if necessary.

The 1989 edition of the code (Section R–210.2, "Emergency Egress Openings") specifies that every sleeping room must have at least one operable window or exterior door that meets requirements for emergency egress or rescue. Such window or windows must be operable from the inside to a full clear opening. The sill height above the floor cannot exceed 44 in. The net clear opening must be at least 5.7 ft². The minimum net clear opening height dimension must be 24 in., and the minimum net clear opening width dimension must be 20 in.

Where a door is used to fill the egress requirements, it must be not less than 3 ft wide and 6 ft 8 in. high.

Smoke Detectors

All homes should be equipped with smoke detectors. If they are not present, they should be installed. There should be at least one detector on each floor of the house. Instruction manuals accompanying the detectors provide information on correct mounting and placement.

Detectors are of no value if they will not function when they are needed. This occurs in a substantial percentage of fatal fires in homes that have detectors.

In a survey taken in 1987, the NAHB found that in 68 percent of fatal fires, there were no smoke detectors in the dwelling. Where they were present, 58 percent were inoperable. Inoperability can be caused by the failure to replace dead batteries and failure to regularly check the functioning of the unit. Homeowners commonly neglect one or both of these requirements during the many years when the detectors are in place and have not been activated.

Older fire detectors that have been in place for some time may not be reliable. Because battery-powered detectors were not introduced until the mid-1970s, average anticipated product life is not known. A 1984 study conducted by the National Fire Protection Association indicated that 16 to 30 percent of the detectors bought in 1974 had stopped working. In rehabilitating a home equipped with smoke detectors that are at least 10 years old, the detectors should be tested frequently and any doubt should be resolved by replacing with new units.

Sprinkler Systems

Smoke detectors can alert occupants to a fire in the house, but they cannot contain or extinguish it. This can be done by a home fire sprinkler system. New types of systems, called fast-response home sprinklers, are designed to respond to a fire more quickly than do standard commercial and industrial systems. Such systems, which are connected to the domestic water supply, can be installed while the house is being rehabilitated.

A piping diagram for a typical home fire sprinkler is shown in Figure 46. The illustrated system includes a water alarm that is set off as the water rises to one or more sprinkler heads when the system is activated by a fire. The occupants receive an audible alarm while the sprinkler operates. The National Fire Protection Association Standard for Home Sprinkler Systems specifies that such an alarm is optional when the system is installed in a home with smoke detectors and mandatory when the system is installed in a home without smoke detectors.

The installation of home fire sprinkler systems in both new housing and as retrofit in existing housing is advocated by the U.S. Fire Administration (USFA). The USFA leaflet, "Home Fire Protection: Quick Response Fire Sprinkler Systems" lists these advantages of installing such systems:

- Small size

    For home systems, sprinklers are smaller than traditional commercial and industrial sprinklers and can be aesthetically coordinated with any room decor.

- Minimal installation work

    When homes are being rehabilitated, installation will require minimal extra piping and labor.

- Low water requirement

    Home systems require less water than industrial and commercial systems and can be fully served by a connection to the existing domestic water supply.

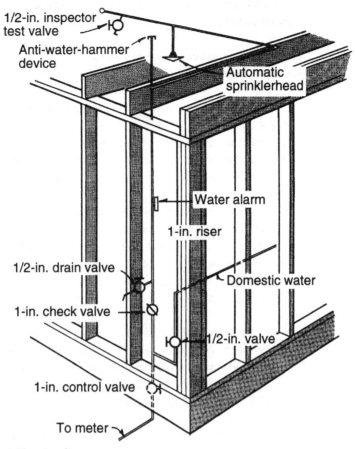

1/2-in. inspector test valve

Anti-water-hammer device

Automatic sprinklerhead

Water alarm

1-in. riser

1/2-in. drain valve

Domestic water

1-in. check valve

1/2-in. valve

1-in. control valve

To meter

Figure 46—Sprinkler pipe diagram.

- Piping requirements

  In addition to metallic pipe, plastic pipe may be used, which has brought down the cost of installation.

Additional features of the systems follow:

- The sprinkler heads react to the temperature in each room, so only the sprinkler heads over the fire will be activated.

- Records kept for a period of 50 years show that it is unlikely that a sprinkler will be activated accidentally.

- Because the quick activation of the system greatly limits a fire's growth, water damage from the system is much less severe than the smoke and fire damage from an unabated fire, and is less than the damage caused by water from fire hoses.

The publication also notes that homeowner insurance underwriters offer discounts for homes equipped with sprinklers.

Additional information on home fire sprinkler systems appears in USFA's publication *An Ounce of Prevention.*

Manufacturers of woodstoves generally recommend that they should be installed professionally. The stoves come with detailed instructions to guide either professional installers or do-it-yourselfers.

A number of jurisdictions require a permit for installing a woodstove and require an inspection after installation.

The Consumer Product Safety Commission provides guidelines for installing woodstoves. They include the following:

- The stove should be placed on a suitable floor protector, such as a brick or concrete hearth.

- Stoves should be at least 3 ft from the walls and ceiling unless the manufacturer's instructions indicate otherwise.

- Chimney connectors should not pass through a combustible wall without proper protection and clearance, as specified by the manufacturer's instructions and local building codes.

Insulated steel pipe should be used for passage through block or masonry, even though these materials are not flammable. An uninsulated steel pipe can raise their temperature sufficiently to bring adjacent combustibles to the flash point. In a horizontal configuration, uninsulated steel flue stovepipe should not be closer than 3 ft to the ceiling.

The Wood Heating Alliance has issued a checklist for factory-built fireplaces:

- Slots, louvers, and other air-circulating components should be clear of all obstructions.

- Stovepipe should never be used in place of a factory-built metal chimney.

- Spacers and supports should not be bent.

- Combustibles and insulation must be set away from the fireplace at least the minimum distance specified in the installation instructions.

- If the fireplace is installed on a combustible floor, a noncombustible safety strip must be placed on the floor beneath the joint between the hearth extension and the fireplace.

- Insulation near the fireplace and chimney should be unfaced and noncombustible. If cellulose-type insulation is used in the attic, the chimney should be properly shielded in accordance with instructions on the insulation package.

- The fireplace should be adequately supported, fastened down, or blocked to prevent it from settling or shifting out of position.

- Trim materials applied to the face of the fireplace should comply with the building code and the manufacturer's instructions. There should not be any cracks between the trim and the fireplace.

- Combustible trim materials, such as mantels, should be located the required distance from the fireplace opening according to the building code and the manufacturer's instructions.

- There should be an airspace between the chimney and combustible materials, as specified in the manufacturer's instructions.

- Chimneys should rise at least 3 ft above the highest point on the roof next to the chimney and at least 2 ft higher than any part of the roof within 10 ft.

There is additional information in "Fireplaces, Woodstoves, and Chimneys" in the "Rehabilitation" section.

Thermal Protection and Moisture Control

Modern comfort requirements and energy costs require that homes be thermally protected by insulation. Tightly constructed and insulated homes also require the proper application of a vapor retarder to prevent or reduce moisture problems. There are many developments continuing in this area. This section will present what is currently known about installing insulation and vapor retarders.

Installment of Insulation

Insulation must be properly installed for good performance. The concept is to develop an "envelope" of insulation around the living space. Insulation can also help control noise and provide additional fire resistance. Installation techniques vary somewhat with different constructions, but the fundamentals of application are essentially the same. Although these techniques are not complicated, certain details are important. Some guidelines and installation tips follow.

General Procedures

All spaces in walls, floors, and ceilings must be carefully fitted with insulation. It should be placed in all small spaces between framing members, between door and window headers and top plates, between jambs and sills, and in the spaces around pipes, wires, and other services that penetrate the top or bottom wall plates and ceilings. If the spaces are small, it may be easier to caulk them rather than to fill them with insulation. If small spaces are not carefully filled, air will pass through them and much of the insulation's effectiveness will be lost. Serious condensation problems can also result.

When there is an offset in framing members such as where ceiling joists join over an interior partition wall, start-and-stop batt and blanket-type insulation at this joint to ensure a tight fit. The insulation will gap and buckle if it is run continuously at the offset. The ends of all batts and blankets should fit snugly. If the insulation is too long it should not be doubled-over or compressed, but cut to fit the space.

Make certain that insulation is placed on the cold (exterior) side of pipes, ducts, wiring, electrical boxes, or other obstructions. Batt and blanket-type insulation may be split and wrapped around the obstruction. If water pipes are located in the outside wall and insulation is not placed between the pipes and the exterior, the pipes may freeze during cold weather. Check local building codes and the manufacturer's recommendations to determine how close a particular insulating material can be installed to a chimney.

Be sure to insulate and weather-strip access panels to attics, crawl spaces, basements, and so forth. Blanket-type insulation can be stapled or glued to the cold side of the access door, and foamed tapes with an adhesive back can be used as weatherstripping.

Exposed vapor retarder facings should be covered because they are somewhat combustible. Breather papers often attached to the reverse side of the batts and blankets are also combustible. Gypsum board, paneling, and ceiling tiles with acceptable flame-spread ratings may be used as a protective covering.

Batt and blanket-type insulation is installed in horizontal and sloped ceilings much like it is in vertical walls. The insulation can be installed from below by stapling before the ceiling finish material is in place, using a pressure-fit blanket, or laying the insulation in from above. The insulation should be extended entirely across the top wall plate, keeping it as close to the plate as possible. If necessary, the gap between the blanket and plate can be stuffed with loose insulation. This will help reduce heat loss and wind penetration beneath the insulation. Take care that the insulation does not block any eave vents. At least 1 in. of clearance should be left between the top of the insulation and the roof sheathing to permit air circulation and moisture evaporation.

If two layers of batt and blanket-type insulation are installed as in a horizontal ceiling area, the top layer should be unfaced or the vapor-retarder facing should be removed. Once the space between ceiling joists is filled with insulation, the second layer should be run at right angles to the first to cover any thermal short circuits caused by framing members. Recessed lighting fixtures should not be covered with insulation because the heat they generate must dissipate to the attic to reduce the fire hazard.

When insulating a sloped or cathedral ceiling, proceed as discussed in the preceding paragraphs. The insulation should extend over the wall plate. Cathedral ceilings are more prone to condensation problems than are conventional attics. Therefore, it is extremely important to minimize air leaks into the roof cavity in this type of ceiling. A continuous airtight vapor retarder without any penetrations is the surest way to ensure this protection. Installing recessed lights in a cathedral ceiling is not recommended because they provide an access point for moist air to enter the ceiling cavity. Local building codes usually require ventilation of the cathedral ceiling cavity. A minimum 3/4-in. airspace should be left between the insulation and the sheathing for that purpose.

In retrofitted attics, the space between the collar beams should be insulated first. The insulation between the rafters should then be fitted snugly against the insulation between the collar beams. The knee wall, if present, must also be insulated. An airtight continuous vapor retarder installed on the warm side of the insulation will minimize the danger of ice dams and condensation on the sheathing. Be certain that the void spaces behind the knee walls and above the collar beams are ventilated.

Another common way to insulate attic ceilings is to install blown-in insulation made of mineral wool or cellulose fiber. For a given manufacturer's loose fill or blown-in insulation, the installed resistance to heat loss depends on the thickness and weight of the material applied per square foot. For this reason, the current Federal specification for mineral-fiber insulation requires that each bag be labeled to show the minimum thickness, the maximum net coverage, and the minimum weight required per square foot to produce specified R-values. Most manufacturers give instructions on each bag for determining how many bags are required for a given attic area to achieve a specific R-value. These instructions must be followed. The amount of blown-in attic insulation to be installed must not be specified or purchased solely on the basis of thickness.

Baffles or pieces of blanket insulation should be installed at the top of exterior wall plates at the eaves without blocking any eave ventilation. Particular care is required for gable and low-slope hip roofs. Blown-in insulation can then be applied over the remainder of the area. This procedure will ensure that those areas are properly insulated and vented.

Batt and blanket-type insulation is generally used to fill wall cavities if the interior wall is not already in place. The batts or blankets are pushed into the wall cavities, usually beginning at the top plate and working down. The end of the batt should either fit or be cut to fit snugly against the bottom plate. At exterior corners and at the intersections of exterior and interior walls, the insulation should be placed in openings between studs before the exterior sheathing is applied. Batt and blanket-type insulation is usually cut about 1 in. larger so it will fit snugly on all surfaces. However, it should not be compressed. Compressing insulation reduces its effectiveness. If the batts are equipped with a vapor retarder, it should be turned to the interior side of the wall. The flanges should be pulled to fit snugly against the edges of the studs and stapled about every 8 in. Avoid gaps. If the batt or blanket has been cut for a nonstandard space and the flange removed from one side, pull the vapor retarder on the cut side to the stud and staple it in place. Be sure to insulate spaces between conditioned and nonconditioned spaces in multilevel homes such as overhanging cantilevered soffit areas, walls between attached garages or storage rooms and living quarters, and areas above and behind built-in kitchen cabinets.

Wall Insulation
Retrofit

For existing structures, insulating walls is much more difficult because they are already enclosed. If the wall cavities will be opened during rehabilitation, they can be insulated as described previously. If it is not necessary to open the wall cavities, two options exist.

If the house is to be resided, rigid insulation board can be added to the outside and a new siding material placed over it. The rigid insulation must be fitted carefully around all windows, doors, and other openings as well as under eaves and porches and at the bottom of the side wall. Taping the joints between the insulation boards is not recommended. A vapor retarder should be carefully installed on the warm side of the wall in areas with winter design temperatures below 0°F or when local building codes require it.

Wood-frame walls in existing houses can be insulated by pneumatically applying fill-type insulation into each of the stud spaces. This procedure is best done by a contractor equipped for such work. Rental equipment may also be available. The siding just below the top plate and below each window is removed on houses with wood siding. Holes are bored through the sheathing into each stud space and additional holes are made below obstructions in the spaces. The depth of each stud space is determined by using a plumb bob. Insulation is forced under slight pressure through a hose and nozzle into the stud space until it is completely filled. Special care should be taken to insulate spaces around doors and windows and at the intersections of interior partitions and outside walls. The wall cavities should be filled completely. The same method can be used in attic and roof spaces that are not accessible for application of other types of insulation. Stucco, brick, and stone veneer walls can be insulated in a similar manner.

Masonry foundation walls may be insulated by using nominal 1- by 2-in. furring strips or, for colder climates, nominal 2- by 3-in. studs spaced 16 or 24 in. on center on the inner surface of the wall, depending on the thickness and type of wall finish. Masonry wall insulation, nominally 1 in. thick (about R-3), is available without a vapor retarder. Use a polyethylene film (minimum 4 mil) or foil-backed gypsum board as the vapor retarder.

For colder climates, the most cost-effective method of providing more thermal protection for masonry walls is to build a frame wall with nominal 2- by 3-in. studs placed 24 in. on center and with 1- by 3-in. bottom and top plates. To provide a thermal break for the studs, set the frame 1 in. from the foundation wall, nail the top plate to the underside of the joists (or blocking between the joists), and fasten the bottom plate to the concrete floor with masonry nails or power-actuated fasteners. Staple R-11 insulation batts between the studs. Install a vapor retarder over the insulation where recommended or required.

Hollow core masonry walls may be filled with loose-fill insulation, such as water-repellent perlite and vermiculite, as well as foamed-in-place types of insulation.

Floors Over Crawl Spaces and Slab on Grade

Floors over crawl spaces may be insulated in one of two ways. If the crawl space is unvented or has operable vent openings, insulate the foundation walls. If the crawl space is vented with fixed vents, insulate the space between the floor joists.

Crawl spaces should be ventilated except when they are used as the plenum for the heating system. Walls of unvented crawl spaces can be insulated by fastening special rigid insulation board to the inside face of the wall, extending from the ground to the top of the wall. Take special care to make certain that the space between the floor joists at the top of the stem wall is insulated. Follow the manufacturer's instructions for the method and type of adhesive or fasteners to attach the insulation to the wall.

Floors over vented crawl spaces can be insulated with blankets of insulation held in place with specially made wire that friction-fits between the joists or with chicken wire nailed to the bottom of the joists. In areas with a winter design temperature of 0°F or lower, a vapor retarder should be installed between the insulation and the floor above. If the bottom surface of the blanket is at the bottom of the floor joists, separate pieces of blanket insulation should be installed at the band-joist area; the vapor retarder should be installed on the crawl space side. If the insulation needs to be protected from the weather, as in the case of an open crawl space or cantilevered addition, plywood or other coverings should be nailed to the bottom of the joists. Be sure any pipes are covered with insulation or are otherwise protected.

Rigid perimeter insulation should be installed at the edges of slab-on-grade floors. The soil should be excavated to either the bottom edge of the footing or about 1 ft below the soil line. The insulation is glued in place, and the excavation is then backfilled. The insulation must be impervious to both moisture and weathering and should be protected from physical damage if it extends above grade.

Vapor Retarders

The concept of vapor retarders and their importance in preventing condensation that can lead to serious decay, paint peeling, and loss of thermal resistance for the insulation is discussed in "Vapor Retarders" under "Thermal Assessment and Moisture Control" in the "Condition Assessment" section. A properly installed vapor retarder also improves the airtightness of the wall or ceiling by acting as an air barrier. Any air leakage around or through a vapor retarder renders it less effective.

General Procedures

The most common type of vapor retarder is a 4- or 6-mil polyethylene film, which can be purchased in rolls of almost any dimension. The polyethylene is best unrolled across the entire wall area covering all windows, doors, and other openings, and stapled to the studs, top and bottom plate, and around any openings. The openings are carefully cut out after the entire sheet is installed. Be careful that all openings, particularly around door and window frames, are carefully covered and that gaps do not develop around electrical boxes. Take care that the vapor retarder covers the area where a partition wall joins an outside wall (Fig. 47). The vapor retarder should be taped at electrical boxes. Repair any tears.

Foil-backed gypsum board may be used instead of polyethylene sheets as a vapor retarder. Some blanket-type insulation has an aluminum or polyethylene vapor retarder attached to one face, which will also suffice. Kraft paper backing is less effective as an air barrier or vapor retarder.

Certain paints may also be used as a vapor retarder if applied to the inside surface of exterior walls. One major manufacturer of household paints has determined that two coats of its alkyd semigloss interior paint, having a dry film thickness of 2.4 mils, has a vapor permeance of 0.9 perm. However, typical latex paints have a relatively high vapor permeance and will not

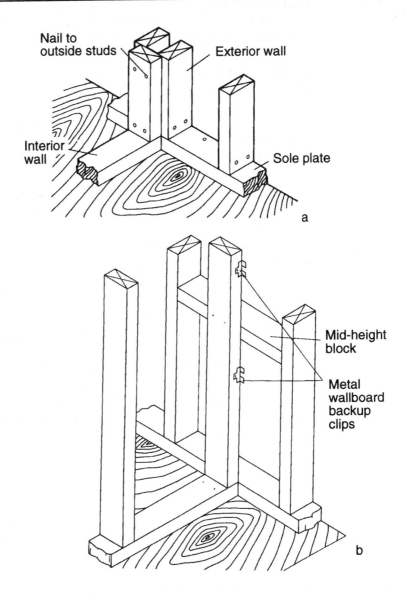

Figure 47—Intersection of interior partition and exterior wall: (a) double studs in exterior wall, (b) horizontal blocking to support partition.

effectively retard the movement of moisture through the wall. The vapor permeance for 4-mil polyethylene is 0.08 perm, and for kraft and asphalt building papers, it is about 0.3 perm. Remember, however, that any gaps bypassing the paint film renders the vapor retarder essentially ineffective.

Blanket insulation with a vapor retarder already attached is often used for vented or unvented crawl spaces and for floors over unheated basements. The retarder is placed upwards towards the winter warm side for cold winter climates. Exposed soil in the crawl space should be covered with a polyethylene vapor retarder. When floor insulation is applied over a partially heated basement or the house is located in a warm, humid climate, the vapor retarder on the insulation may face down.

If winter design temperatures are 0°F or lower, place a vapor retarder on the winter warm side of the ceiling and provide adequate attic ventilation. The vapor retarder must be installed so that there is no gap at the junction of the ceiling and outside wall.

Summer air-cooling for comfort in temperate climates does not usually create serious vapor problems in exterior walls and ceilings. Normally, the cooled air is not much colder than the dewpoint of the outdoor air. Therefore, it is generally best to locate the vapor retarder for the more serious case of winter condensation and to disregard the summer case, even though there may be occasional condensation.

Humid climates as well as cold climates (Fig. 48) deserve special consideration.

Special Considerations for Humid Summer Climates

The American Society of Heating, Refrigerating and Air-Conditioning Engineers (ASHRAE) has defined a humid climate as follows:

- A 67°F or higher wet-bulb temperature for 3,500 hours or more during the warmest 6 consecutive months of the year, or

- A 73°F or higher wet-bulb temperature for 1,750 hours or more during the warmest 6 consecutive months of the year. (For practical purposes, wet-bulb temperatures and dewpoint temperatures can be assumed to be approximately the same.)

If dwellings are constantly air-conditioned in these areas, warm moist air can move from the outside and condense on the cooler inside portion of outside walls. This situation is the reverse of cold weather condensation in colder climates, but the same principle applies. The amount of probable condensation is difficult to generalize and is dependent upon local conditions. Therefore, successful, trouble-free local practices should be followed. In general, exterior surfaces of the building should be airtight and more vapor-resistant than interior surfaces. Therefore, an interior vapor retarder is not recommended.

If vapor retarders are used, they should be on the outside. Any water that does enter the outside surface of the structure can flow through to the inside where it can be removed by the air-conditioning system instead of accumulating in the floor, wall, or roof construction.

In addition, the cooling system should be properly designed and be able to adequately dehumidify the incoming air without over-cooling. Proper maintenance and operation of the air-conditioning system is important.

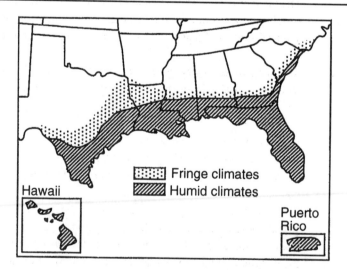

Figure 48—Humid climates in the continental United States follow the coastal belt of the Gulf of Mexico and South Atlantic coastal areas. Reprinted with permission from the American Society of Heating, Refrigerating and Air-Conditioning Engineers.

| Indoor Humidity Control | The manner in which a home is occupied and the occupant's lifestyle can greatly affect the relative humidity level and, thus, the potential for condensation during cold weather. A family of four may produce as much as 2 to 3 gal of water vapor per day. About half of this is due to the moisture exhaled from the body in the normal breathing process. The other half is a result of showering, bathing, cooking, washing dishes, washing clothes, and similar water-consuming tasks. House plants are also a source of moisture. Humidifiers are, of course, an intentional source of humidity during the winter. Water vapor from these activities and sources increases the indoor relative humidity. |

Moisture released from many activities cannot easily be reduced or eliminated. However, some measures can be taken. Vent clothes dryers, bathrooms, and kitchens to the outside, and use the vents regularly. Use lids when cooking, and do not hang laundry inside to dry. In cold climates, do not set humidifiers at a level higher than 40 to 45 percent relative humidity.

Moisture can also come from all the sources discussed in the section on wood decay ("Vulnerable Areas" under "Recognizing Damage" in the "Condition Assessment" section).

The higher the relative humidity in a house, the greater the chance that some surface will be cool enough (even in moderate or warm climates) to cause condensation. For this reason, keeping indoor relative humidity below about 45 percent, if possible, is a good way to avoid many moisture problems that may be related to constantly high indoor relative humidity. In many areas with a cold winter climate, humidity can be controlled with exhaust fans, balanced ventilation systems, or heat recovery ventilators. The choice of equipment depends on other factors outside the scope of this manual. In mild, humid climates (for example, coastal), dehumidifiers may be an efficient and effective way to remove moisture. Ventilation is less effective in such climates.

Site Improvements

Various site conditions can greatly affect the useability and desirability of a home. Grading the site for proper drainage is critical, as is the proper use of retaining walls. Other factors to consider include clearance between the wood members and the soil line, and the presence or absence of trees, walkways, driveways, patios, and fencing. Suggestions for correction of possible problems are provided here.

Grading

The lot should be graded so that all water drains away from the house. If there are gutters, make certain that water does not accumulate or that the soil does not wash away where the downspouts discharge. Splash blocks will help direct water and prevent erosion.

Retaining Walls

Retaining walls may be used on hillsides or other difficult sites. Retaining walls help to increase available level space or divert water, particularly on steep hillsides. Be sure to check local regulations before beginning construction.

Pressure-treated rectangular wood timbers or railroad ties may be used to construct retaining walls (Fig. 49). The timbers are stacked so that the butted ends of the members in one course are offset from the butted ends of the members in the courses above and below. The bottom course should be placed at the base of a level trench. In well-drained sandy soil, there is no need for special footing preparation and materials. In less well-drained soils, 12 to 24 in. of gravel backfill behind the wall and a 6-in.-deep gravel footing are desirable. Each course of timbers should be nailed to the course below, using galvanized spikes with lengths 1-1/2 times the thickness of the timbers. Every other course of timbers should include members inserted perpendicularly to the face of the wall and nailed with spikes to the lower course. These tieback members should extend horizontally into the soil behind the wall for a distance equal to their distance above the base of the wall. The end of the tieback member should be nailed to a "deadman" timber 24 in. long that has been buried horizontally in the soil and aligned parallel

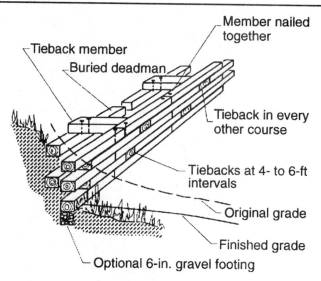

Member nailed together

Tieback member

Buried deadman

Tieback in every other course

Tiebacks at 4- to 6-ft intervals

Original grade

Finished grade

Optional 6-in. gravel footing

Figure 49—Pressure-treated timber retaining wall.

to the timbers in the wall. These tiebacks and deadmen should be installed in every other course at intervals of 4 to 6 ft along the retaining wall. The tiebacks and deadmen in a course should be located midway between those in the second course below. The objective of the deadmen and tiebacks is to prevent the finished wall from tipping over from the pressure of the soil retained by the wall.

Retaining walls can be constructed solely of poured concrete (Fig. 50). The footing for the wall is dug to a depth below the frost line. Then a form is built in which to pour the concrete for the footing and wall as a single unit. The form for the face of the wall should be vertical, but the back of the wall should be built at an angle to provide a wall that is thicker at the base. Reinforcing rods (5/8-in.) spaced on 12-in. centers should be placed in the form and wired together to form a lattice. Concrete is poured in the form to the depth of the footing and allowed to set partially before the vertical portion of the wall is poured. Backfilling the wall with 12 to 24 in. of gravel is recommended.

Clearance

If any wood is closer than about 8 in. to the soil line, it will be necessary to regrade or remove soil. A wheelbarrow and shovel may suffice for small amounts, but a small mechanical loader may be needed for a larger job. Make certain that enough soil is removed so that the water will drain away from the foundation. Do not slope the soil toward the foundation. In areas with termites, remove earth-filled flower planters that are attached directly to the house because they provide an excellent avenue for termites to enter the structure.

Soil Condition

During construction, the top soil is often removed or buried under an inferior soil. Heavy equipment can also cause soil compaction. Various building materials such as plaster or cement contain lime, which, if they are buried, can increase the alkalinity (pH) of the soil to an excessive level. As a result, shrubbery, trees, grass, or other herbaceous plants material may appear to be doing poorly. If plans call for new plants to be established, have the soil tested for pH and nutrients. The Cooperative Extension Service, which is part of the State land grant institution and has offices in nearly every county, will conduct the tests for a nominal fee. They frequently have horticulturists and horticultural publications available.

Trees

Normally, trees do not cause problems. They can greatly improve the value and quality of a homesite. However, when they are planted too close to foundations, driveways, or walkways, or when volunteer trees establish themselves in these locations, the roots can crack foundations

100

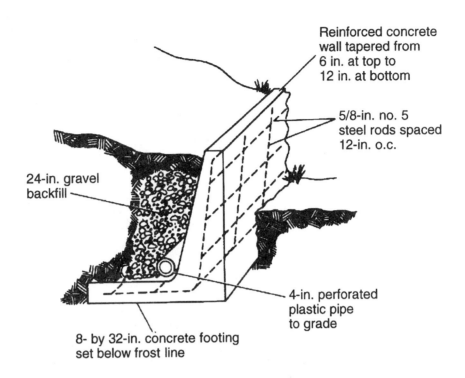

Reinforced concrete wall tapered from 6 in. at top to 12 in. at bottom

5/8-in. no. 5 steel rods spaced 12-in. o.c.

24-in. gravel backfill

4-in. perforated plastic pipe to grade

8- by 32-in. concrete footing set below frost line

Figure 50—Reinforced concrete retaining wall (maximum 4 ft high above grade).

and cause driveways and walkways to bulge upward. The trees, of course, can be removed. To ensure that the stump does not sprout and continue growing, apply a herbicide to the cut stump to kill the root system. Local farm or garden stores can supply the appropriate herbicide, or a landscape architect may be consulted.

House Moving

Lastly, if the site on which the house is located is not satisfactory or cannot be adequately improved, consider moving the house. Such a move is an expensive and involved operation. Permits to use the public road system are usually required. The structure must be low enough to clear any overhead utility lines; if not, representatives from these companies will need to be present to move the lines. The original and receiving sites must be level enough so the structure can be moved from its old location to its new one. Bridges must be wide enough and without overhead obstructions that would prevent the structure from passing. Obviously, the shorter the distance the structure is moved, the less trouble and expense is involved. House moving requires the services of a professional, who should be consulted before the structure is purchased or before much consideration is given to moving the house.

Major Structural Features

Foundations and Basements

A sound foundation is a prime requisite of a good house and a house worth rehabilitating. Foundations for residential construction can be divided into three general categories: basements, crawl-space-type homes, or pier-and-beam-type construction (either enclosed with a perimeter wall or open) and concrete slabs. The basic features of each will be discussed in order to provide an understanding of what constitutes good construction practices and what might be involved in any needed repair.

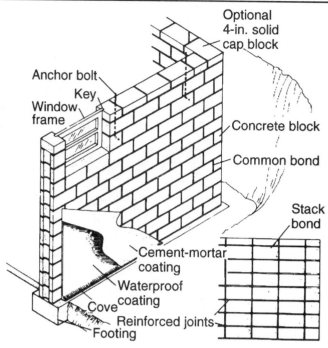

Figure 51—Concrete block foundation wall.

Figure 51 shows the different elements of a concrete block foundation wall.

The footings act as the base of the foundation wall and transmit the weight of the foundation wall and house to the soil. Footings are nearly always poured concrete, although in some older homes, large cut stones or brick may have been used. The type and size of footings should be suitable for the soil condition. In cold climates, the footings should be far enough below finished grade level to be protected from frost. Local codes usually establish this depth, which is often 4 ft or more in the northern United States and in Canada.

Foundation walls form the enclosure for the basement and carry the wall, floor, roof, and other building loads. Foundation walls may be concrete block or poured concrete. Poured concrete foundation walls may be keyed into the footing (Fig. 52). In older homes, brick, field stone, cut limestone, or other natural materials may have been used. Foundation walls at least 7 ft 4 in. high are desirable for full basements; 8-ft walls are commonly used. Preservative-treated wood is being used for basements in some newer homes.

Shallower basements are common in many older homes. In some cases, the depth of these basements can be extended by additional excavation, if the footings are deep enough.

Poured concrete walls can be damp-proofed with one heavy cold or hot coat of tar or asphalt. The coating should be applied to the outside from the footings to the finish gradeline when the surface of the concrete has dried enough to assure good adhesion. Such coatings are usually sufficient to make a wall watertight against ordinary seepage that may occur after a rainstorm. In addition, the backfill around the outside of the wall should be gravel. A gravel backfill prevents soil from holding water against the foundation wall and allows the water to flow quickly down to the drain tile at the base of the wall. Instead of gravel backfill, a drainboard composed of plastic fibers or polystyrene beads can be installed against the foundation wall, providing the same function as the gravel backfill. In poorly drained soils, a membrane instead of a coating may be necessary.

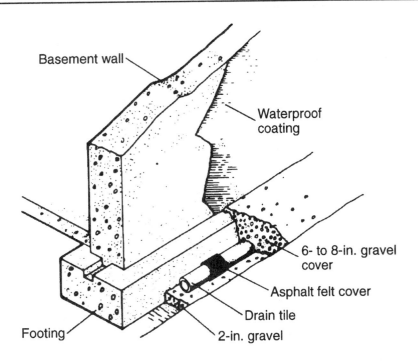

Basement wall

Waterproof coating

6- to 8-in. gravel cover

Asphalt felt cover

Drain tile

2-in. gravel

Footing

Figure 52—Poured concrete basement wall keyed to footing. Proper installation of basement waterproofing and drain tile are also shown.

Concrete block walls may be waterproofed by applying a coating of cement mortar over the block with a cove formed at the juncture with the footing (Fig. 51). When the mortar is dry, a coating of asphalt or other waterproofing normally will assure a dry basement. Other methods include the application of a waterproof membrane, such as a 6-mil polyethylene film over the asphalt, to provide a water barrier.

For the basement floor, a base of 4 in. of compacted gravel is normally laid over the soil. A 6-mil polyethylene vapor retarder is placed over the gravel and a 4-in.-thick concrete slab is poured on top. Any interior drain pipes, plumbing, or such must be laid in place before the concrete is poured. At least one floor drain is normally installed near the laundry area.

Drains are often used around the exterior of foundation footings enclosing basements, around habitable spaces below the outside finish grade (Fig. 52), in sloping or low areas, or in any location where it is necessary to drain away subsurface water. This can help prevent damp basements and wet floors.

Drains are installed at or below the area to be protected, and should drain toward a ditch or into a sump where the water can be pumped to a storm sewer. Perforated plastic drainpipe, 4 in. in diameter, is ordinarily placed at the bottom of the footing level and on top of a 2-in. gravel bed (Fig. 52). Another 6 to 8 in. of gravel is placed over the pipe. In some cases, 12-in.-long clay tile is used to form the drain. Tiles are spaced about 1/8 in. apart, and the joints are covered with a strip of asphalt felt. Drainage is toward the outfall or ditch. Dry wells for drain water are used only when the soil conditions are favorable for this method of disposal. Consult local building regulations before constructing such a system.

Crawl Spaces

The foundation for a crawl-space-type home is constructed in a manner similar to that for a full basement, except that a basement floor is not needed and the foundation wall is only tall enough to provide at least 18 in. of clearance between the bottom of the floor joists and the interior finish grade. This clearance allows room for plumbing, electrical, and other services to be installed and maintained. With adequate venting, the clearance also provides good air circulation and helps keep the wood dry. A polyethylene vapor retarder should be applied over the soil.

For larger houses, piers to support the floor system are installed within the perimeter of the foundation walls. Piers have footings just as foundation walls do, and the upper part may be made of poured concrete or block. In mild climates, footings are located only slightly below the finish grade. However, in the northern states where frost penetrates deeply, the footing is often located 4 ft or more below the finish grade.

In the southern United States where mild weather conditions exist most of the year, a perimeter wall is not always used. This cuts construction costs and allows ample ventilation. However, it also provides easy access to the underside of the house by unwanted animals and other pests. If the perimeter of one of these homes is to be enclosed, ample room should be allowed for ventilation and the soil should be covered with a vapor retarder.

Concrete Slabs

Figure 53 shows a combined floor slab and footing foundation system. This type of construction is also called a thickened edge or monolithic slab and is common in warm climates where frost in the soil is not a problem. The bottom of the perimeter footing should be at least 1 ft below the grade line.

Sloping ground or low areas are usually not ideal for slab-on-grade construction because structural and drainage problems can add to costs. However, split-level houses often have a portion of the foundation designed for slab-on-grade. In such instances, the slope of the lot is taken into account and can become an advantage.

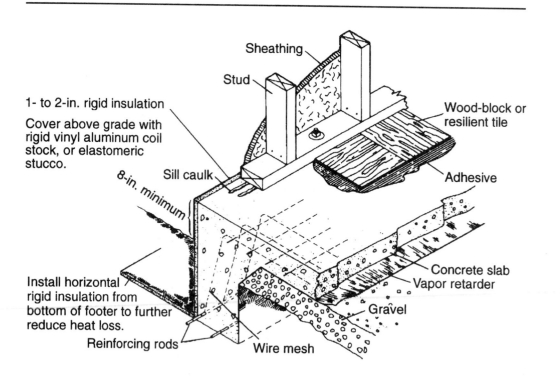

Figure 53—Combined floor slab and footing foundation system.

Concrete slabs can also be poured directly on the ground. The top of the slab should be at least 8 in. above the ground. Other requirements are similar to those for the combined floor slab and footing foundation system.

Independent concrete slab and foundation walls (Fig. 54) are suitable in climates where the ground freezes to an appreciable depth. The walls of the house must be supported by foundation walls or piers that extend below the frost line to solid bearing on unfilled soil. In such construction, the concrete slab and the foundation wall are usually separate.

Correction of
Moisture Problems

Basements are common in houses located in the northern states. They provide room for storage, recreational areas, and workshops, and in many situations are at least partially converted into living space. Unfortunately, improper construction or location of the house on a particular soil type can cause moisture problems. These problems can range from condensation to seepage, leakage, or other problems.

Condensation is the most common basement moisture problem and is the most easily corrected. It results when warm air, which is capable of holding a greater amount of water vapor than is cooler air, comes in contact with colder surfaces such as basement walls, floors, and cold water pipes. As the warmer air cools, the excess water vapor is deposited on these surfaces in the form of condensation.

If condensation is the sole source of moisture, a number of relatively easy steps may be taken to control it. First, reduce or eliminate moisture-producing activities throughout the house, especially in the basement, such as hanging clothes on a line to dry, using an unvented clothes dryer, showering without using a vent fan, and storing "green" firewood. Basement windows should be kept closed in the summer when it is warmer outside than in the basement. Cold water pipes should be insulated. If these remedies are inadequate, consider using a dehumidifier.

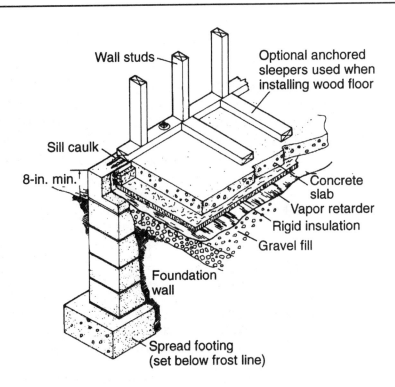

Figure 54—Independent concrete slab and foundation wall system for deep frost-line climates.

For year-round condensation control as well as reduced heating and cooling costs, insulate and install a vapor retarder over basement walls as described in "Retrofitting Wall Insulation" under "Thermal Protection and Moisture Control" in the "Rehabilitation" section.

Seepage is also fairly common. It results from surface moisture or soil saturation and may be moderately easy and inexpensive to correct.

Seepage results when soil becomes wet and perhaps saturated from runoff of roof rainwater (absence of gutters or inadequate downspout leaders), soil settles along the basement wall, a positive grade away from the foundation does not exist, rainwater collects in window wells, or lawn or shrubs next to the foundation wall have been watered excessively. This problem may be compounded by porous masonry walls, deteriorated joints, or cracks.

Relatively small or moderate amounts of water may seep into the basement. Indicators of seepage are standing water along the outside of the foundation wall following rain or snow melt; localized areas of wetness on the floor or wall, indicating capillary transfer from saturated soil; or seepage at cracks or joints.

Seepage can often be corrected by making certain that all water runs away from the house. Be sure to check the soil slope as well as the slope of driveways, porches, patios, and such. Downspouts should discharge roof water at least 36 in. from the foundation. If the ground slopes toward the foundation, some added fill may be helpful, but be certain to maintain at least 8 in. of clearance between any wood and the soil line. If termites are prevalent, it may be necessary to treat the new soil. Covers can be installed over window wells to protect them from moisture.

If these remedies do not correct the problem and seepage continues, test two coats of a water-proofing paint or compound on a basement wall problem area. If the test is successful after several weeks of trial, cover the entire area. All holes and cracks must be patched before applying the waterproofing.

When seepage is entering under pressure from the outside, a weep pipe should be installed at the floor-wall junction where the pressure is greatest. After chiseling a dovetail groove, start patching at the top and finish at the bottom of the crack. If water continues to be discharged through the weep pipe, leave the pipe in and divert the discharge to a sump or drain by means of a hose. If the weep pipe is to be removed, replace it with a cone-shaped plug of putty-like-consistency hydraulic cement or mortar mix. If this plug fails to prevent seepage, reinstall the weep pipe and refer to the following information on leakage.

If seepage is occurring at the floor-wall junction, a double thickness of waterproofing compound may be used to seal the joint if the seepage is slight. If the seepage is moderate, a 2-in.-wide by 1-in.-thick dovetail groove must be chiseled out and brushed clean before hydraulic or mortar cement is used to seal the joint. If the seepage is heavy or under pressure, weep pipes should be installed to discharge the water through hoses or concrete troughs should be designed to carry it to a sump or drain.

Leakage is the least common basement moisture problem and the most difficult and costly to correct. It is most common in dense clay or silt soils that form an impervious layer that prevents moisture from draining away from the foundation. However, leakage may occur in any soil type, especially near marshes or in hilly areas where the ground water is high and has been intercepted by the foundation.

With leakage, large amounts of water will enter the basement, generally under some degree of pressure, through cracks in the wall, floor, or floor-wall joints. In severe situations, the concrete walls or floor may actually buckle and rupture. If a high ground-water level or impervious soil problem is suspected, the local Natural Resources Conservation Service office may provide diagnostic assistance and referral. Frequently, neighbors with similar basements will have experienced moisture leakage and can give advice about successful remedies.

Leakage problems are not easily corrected, and therefore are expensive as well as messy and time-consuming. In moderately severe situations caused by a high water table, it may be possible to relieve water pressure with a weep pipe baseboard sump pump(s) system. In this situation, the sump is surrounded by a layer of clean gravel and holes are drilled into upper portions of the sump so water can drain into it and relieve under-floor pressure.

In severe situations, a combination of drain tile installed under the floor at the perimeter of the interior foundation wall with weep pipes (floor must be broken up) and a sump pump may remedy the problem.

In extremely severe situations, it will be necessary to excavate around the exterior of the foundation, seal it with a continuous waterproof membrane (Fig. 52), and install exterior drain tile to a sump. In northern climates, a homeowner facing this unpleasant task should also consider insulating the exterior of the foundation wall before backfilling. If this does not relieve floor moisture conditions, the floor may need to be repoured over a continuous waterproof membrane.

Insect Control

As discussed in "Insects" under "Recognizing Damage" in the "Condition Assessment" section, subterranean termites range throughout much of the United States. New construction in termite-prone areas is treated with insecticides to prevent infestation. Existing construction can also be effectively treated. If any changes are made to the foundation of the existing structure or the soil level is changed, treatment for termites is recommended. Treating both new and old construction requires the use of insecticides, many of which can be purchased only by specially trained and certified applicators. If not properly applied, these pesticides can contaminate water systems, heat ducts, and other parts of the house. Since special knowledge and equipment are required to effectively use these insecticides, it is probably best to hire a pest control contractor.

Cracks in Concrete

Minor hairline cracks frequently occur in concrete walls during the curing process, but usually require no repair. Open cracks should be repaired, but the type of repair depends on whether the crack is active or dormant and whether waterproofing is necessary. One of the simplest methods to determine if the crack is active is to place a mark at each end of the crack and observe whether the crack grows beyond the marks.

If the crack is dormant, it can be repaired by routing and sealing. Routing is accomplished by following along the crack with a concrete saw or chipping with hand tools to enlarge the crack. The crack is first routed 1/4 in. or more in width and to about the same depth. Then the routed joint is rinsed clean and allowed to dry. A joint sealer, such as an epoxy cement compound, should be applied in accordance with manufacturer's instructions.

Active cracks require an elastic sealant, which should be applied according to the manufacturer's instructions. Sealants vary greatly in elasticity; use a good-quality sealant that will remain pliable. The minimum depth of routing for these sealants is 3/4 to 1 in., and the width is about the same. The elastic material can then deform with movement at the crack. Strip sealants can also be applied to the surface, but they tend to protrude and may be objectionable.

Crumbling Mortar

Crumbling mortar joints in masonry foundations or piers should be repaired. First, chip out all loose mortar and brush the area thoroughly to remove all dust and loose particles. Before applying new mortar, dampen the clean surface so that it will not absorb water from the mixture. Premixed mortar can be purchased. It should have the consistency of putty and should be applied like a caulking material. For a good bond, force the mortar into the crack to contact all depressions, then smooth the surface with a trowel. Protect the surface from sun and wind for a few days, and wet the surface occasionally to keep the mortar from drying out too rapidly. For best results, the crack should be enlarged in the shape of a dovetail, which will help to lock in the waterproofing sealer.

Uneven Settlement

Uneven settlement in a concrete foundation caused by poor footings or no footings at all usually damages the foundation beyond repair. In a pier foundation, the individual pier or piers may be replaced. If the pier has stopped settling, blocking may be added on top of the pier to level the house. In either situation, the girder or joists supported by the pier must be jacked and held in a level position while the repairs are made.

Floor Systems

The floor system in many older homes consists of columns or posts, beams or girders, sill plates, joists, and subfloor (Fig. 55). Assembled on a foundation, this system forms a level, anchored platform for the rest of the house and a strong diaphragm to keep the lateral earth pressure from pushing in the top of the foundation wall. The columns or posts and the beams of wood or steel that support the joists over a basement are sometimes replaced with frame or masonry walls when the basement area is divided into rooms. Second-story floors are generally supported on load-bearing walls that, in turn, are supported by the platform. Wood-frame houses may also be constructed over a crawl space with floor framing similar to that used over a basement, or they may be constructed on a concrete slab. Dry lumber should always be used to avoid settlement problems and warping caused by shrinkage; the recommended maximum moisture content is 15 percent.

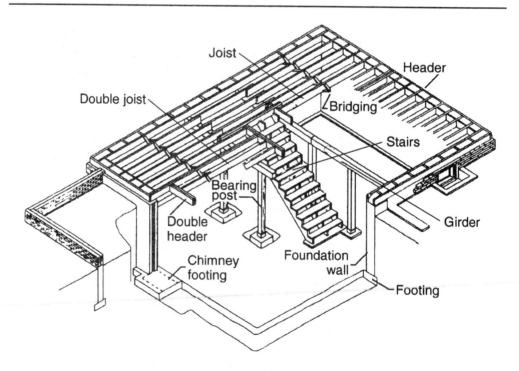

Figure 55—Framing for floor system and basement.

108

Basement Posts   Any type of basement post may settle because of inadequate footings. Wood posts may deteriorate and settle because of decay or insect damage. To correct either problem, a well-supported jack must be used to raise the floor girder off the post in question. This re-leveling must be done slowly and carefully to avoid cracking the plaster in the house walls. A steel jack post is a convenient replacement for the removed post.

Methods for installing jack posts are shown in Figures 56 and 57. When jack posts are used to stiffen a floor or to carry light loads, they can be set directly on the concrete floor slab. If they are used to support heavy loads, a steel plate may be necessary to distribute the load over a larger area of the floor slab. The jack post should not be used for heavy jacking. If heavy jacking is required, use a regular jack to lift the load carefully and then put the jack post in place.

Jacking houses involves very large loads. Be sure that the jack and its support system are level and perpendicular to the load to be lifted. Otherwise, the jack or support system can pop out of place, possibly causing personal injury or damage to the structure.

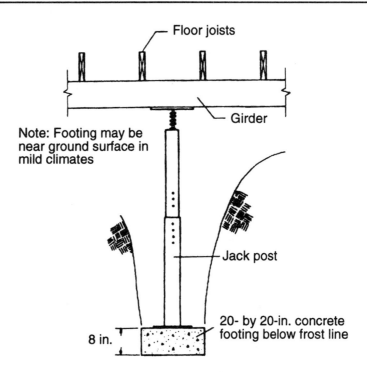

Figure 56—Jack post supporting sagging girder in crawl space.

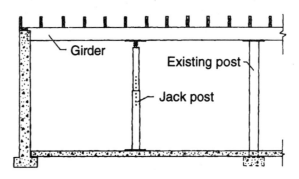

Figure 57—Jack post used to level sagging girder in basement.

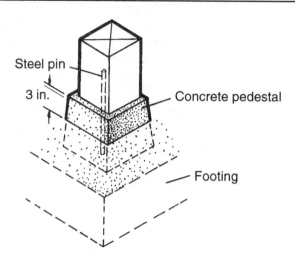

Steel pin

3 in.

Concrete pedestal

Footing

Figure 58—Basement post on pedestal above floor.

If a wood post is used, a pedestal should be built to raise the base of the post slightly above the floor surface (Fig. 58). This will allow the end of the post to dry out if it becomes wet so that it is less subject to decay.

Posts that have been pressure-treated with a wood preservative may be set directly on the concrete. Cutting can expose untreated wood. If the end of the treated post must be cut off, the cut end should be turned up. The post should be firmly attached at both the top and bottom.

Replacement of Framing Members

If an examination of the floor framing reveals decay or insect damage in a limited number of framing members, replace the affected members or repair the affected sections. Replace damaged members with preservative-treated wood if exposure conditions are severe. Large-scale damage may cause the house to be classified as not worth rehabilitating.

Replacement requires that the framing supported by the damaged member be temporarily supported by jacks in a crawl space or jacks with blocking in a basement. A heavy crossarm on top of the jack will support a 4- to 6-ft width. If additional support is necessary, more jacks are required. Raise the jacks carefully and slowly and only enough to take the weight off the member to be removed. Excessive jacking can crack plaster and pull the building frame out of square. After the new or repaired members are in place, gradually take the weight off the jack and remove it.

Sometimes there is decay or termite damage in only a small part of a member. An example might be the end of a floor joist supported on a concrete foundation wall. Such a joist may be decayed only where the wood contacts the concrete. After applying a brushed-on preservative to the affected area, jack the existing joist into place and nail a short length of new material to the side of the joist (Fig. 59). If the condition that caused the deterioration is not corrected, the new material should be pressure-treated.

Floor Leveling

After the repairs to the principal supports of the first floor platform have been properly made, the support points for the floor should be level; however, the floor may still sag. If the floor joists have sagged excessively, permanent set may have occurred and little can be done except to replace the floor joists. A slight sag can be overcome by nailing a new joist alongside alternate joists in the affected area. If the new joists are slightly crooked, place them with the crown (crook) up and force them into alignment with the sagging joists. Joists that are not

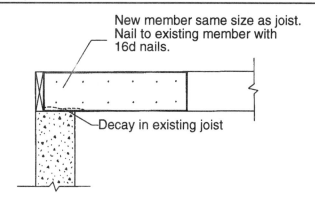

New member same size as joist.
Nail to existing member with
16d nails.

Decay in existing joist

Figure 59—Repair of joist with decayed end contacting the foundation.

crooked should be inspected for the presence of knots along the edge. The largest edge-knots should be placed on top because knots on the upper edge of a joist are placed in compression and will have less effect on joist strength. Each new joist must be jacked at both ends to force the ends to the same elevation as the existing joist. The same treatment can be used to stiffen springy floors.

Girders that sag excessively should be replaced. Excessive set cannot be corrected. Jack posts can be used to level slightly sagged girders or to install intermediate girders, but unless the space is used very little or jack posts can be incorporated into a wall, the added posts are generally in the way. The best solution may be to jack up the supported members, remove the girder, and replace it.

If solid timbers are not available for girders, built-up beams may be constructed by nailing three or four layers of dimension lumber together. The built-up beam may be made longer than any of the individual members by butt-jointing members together. These butt joints must be staggered between adjacent layers so that they are separated by at least 16 in. In addition, the built-up beam must be supported by a column or pier positioned within 12 in. of the butt joints (Fig. 60). The ends of wood beams should bear at least 4 in. on the masonry walls or pilasters. If wood is untreated, a 1/2-in. airspace should be provided at each end and side where the beam must contact masonry (Fig. 60). The top of the beam should be level with the top of the sill plates on the foundation walls. Using a wood plate over wood beams is not necessary because floor joists may be nailed directly to the beam.

In a home with a crawl space, do not reinforce a weak floor system by propping up joists with wood supports. Wood, as well as masonry, provides an easy pathway for termites to enter the structural framing. If new masonry supports (piers) are added, be sure to treat the soil under and around them to prevent termite infestations.

Elimination of Squeaks

One of the most common causes of squeaking is inadequate nailing. To correct this condition, drive a nail through the face of the flooring board near the tongued edge into the subfloor—preferably also into a joist. Set the nail and fill the hole. A less objectionable method from the standpoint of appearance is to work from under the floor using screws turned through the subfloor into the finish floor. This will also bring warped flooring into a flat position.

Squeaks in flooring frequently are caused by movement of the tongue of one flooring strip in the groove of the adjacent strip. One of the simplest remedies is to apply a limited amount of mineral oil to the joints.

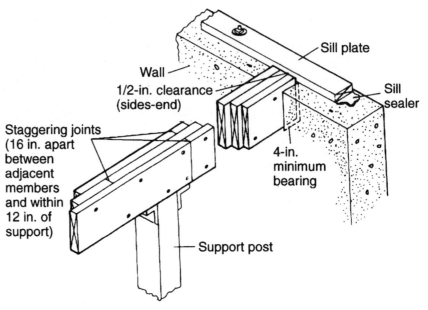

Figure 60—Typical built-up wood beam installation.

Strip flooring installed parallel to the joists may also deflect excessively. Solid blocking nailed between joists and fitted snugly against the subfloor (Fig. 61) will prevent this deflection if it is installed at relatively close spacing.

Sagging floor joists often pull away from the subfloor and cause excessive deflection of the floor. If this is the cause of squeaks, squeeze construction mastic into the opening between the subfoor and joists. An alternate remedy is to drive small wedges into the spaces between joists and subfloor (Fig. 62). Drive them only far enough for a snug fit. This method of repair should be limited to a small area.

Undersized floor joists that deflect excessively are also a major cause of squeaks. Adding girders (described previously) to shorten the joist span is the best solution to the problem.

Framing for Openings, Projections, and Bathtubs

In many older homes, it is sometimes desirable to make relatively important structural modifications. The modifications may include openings in the floor system to install a new set of stairs, installation of fireplaces or chimneys, addition of a bay window, or installation of additional bathtubs. Exercise care in making these modifications.

Large openings in the floor, such as stairwells and fireplaces or chimneys, usually interrupt one or more joists. Such openings should be planned so that their long dimension is parallel with the joists to minimize the number of joists that are interrupted. If possible, the opening should be coordinated with the normal joist spacing on at least one side to avoid the necessity for an additional trimmer joist to frame the opening.

A single header is generally adequate for openings up to 4 ft wide. A single trimmer joist at each side of the opening is usually adequate to support single headers that are located within 4 ft of the end of joist spans (Fig. 63). Tail joists under 6 ft in length may be fastened to the header with three 16d end nails and two 10d toe nails or their equivalent. Tail joists over 6 ft long should be attached with joist hangers. The header should be connected to trimmer joists in the same manner that tail joists are connected to the header.

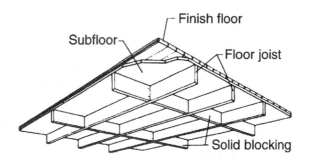

Figure 61—Solid blocking between floor joists where finish floor is laid parallel to joists.

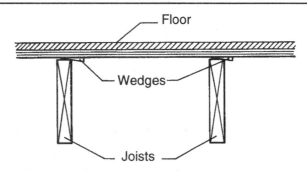

Figure 62—Wedges driven between joists and subfloor to stop squeaks.

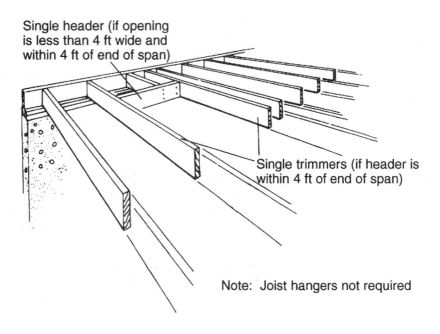

Figure 63—Floor opening framed with single header and single trimmer joists.

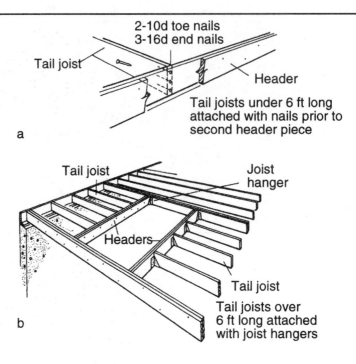

Figure 64—Floor opening framed with double header and double trimmer joists: (a) nailing tail joists under 6 ft long; (b) joist hangers used for longer tail joists.

If wider openings are unavoidable, double headers are generally adequate up to 10 ft (Fig. 64). Tail joists may be connected to double headers as specified previously for single headers. Tail joists that are end-nailed to a double header should be nailed before the second member of the double header is installed so that the nails penetrate adequately into the tail joists. A double header should always be attached to the trimmer with a joist hanger.

Trimmer joists at floor openings must be designed to support the concentrated loads imposed by headers. As noted previously, a single trimmer is adequate to support a single header located near the end of the span. All other trimmers should be at least doubled and should be engineered for specific conditions.

The framing for wall projections such as bay windows, wood chimney enclosures, or first- or second-floor extensions beyond the lower wall should consist of the projection of the floor joists (Fig. 65a).

Extensions normally should not exceed 24 in. Balconies extending beyond 24 in. need special design consideration. The subflooring is carried to and sawed flush with the outer framing member of the balcony. Greater projections for special designs may require special anchorage at the opposite ends of the joists.

Projections at right angles to the length of the floor joists should be limited to small areas and extensions of not more than 24 in. If the projecting wall carries any significant load, the joists should be doubled (Fig. 65b). Joist hangers should be used at the ends of the members.

A bathtub full of water is heavy and may cause floor joists to deflect excessively. A doubled floor joist should be installed beneath the tub to support this load (Fig. 66). The intermediate joist should be spaced to allow installation of the drain. Use metal hangers or wood blocking to support the edge of the tub at the wall.

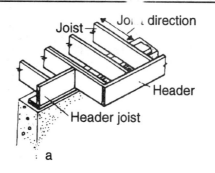

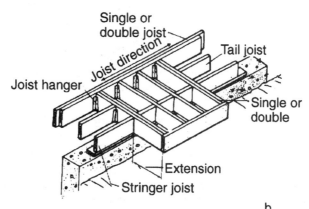

Figure 65—Floor framing at wall projections: (a) continuing of floor joists, (b) projection perpendicular to floor joists.

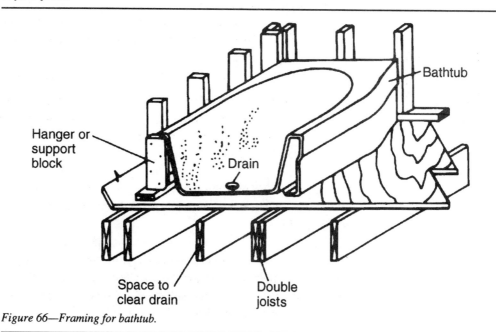

Figure 66—Framing for bathtub.

Cutting of Floor Joists

Although cutting, notching, or drilling joists should be avoided, it is sometimes necessary to conceal plumbing pipes or wiring in a floor. Joists or other structural members that have been cut or notched can sometimes be reinforced by nailing a reinforcing scab to each side or by adding an additional member.

Notching the top or bottom of the joist should be done only in the end one-third of the span and to not more than one-sixth of the joist depth. If greater alterations are required, headers and tail joists should be added around the altered area similar to the method used at a stair opening (Fig. 63).

When necessary, holes may be bored in joists if the diameter is no greater than one-third of the joist depth and the edge of the hole is at least 2 in. from the top or bottom edge of the joist (Fig. 67). If a joist must be cut and these conditions cannot be met, double the joist or reinforce it by nailing scabs to each side.

New Floor Covering

Floor coverings are available in a variety of materials: wood in various forms; asphalt, vinyl, rubber, and cork tile; ceramics; linoleum; sheet vinyl; carpeting; and liquid seamless flooring. The material selected depends on existing conditions, the planned use of the floor, and the homeowner's budget.

Before any floor is laid, a suitable base must be prepared. Unless existing wood flooring is exceptionally smooth, it should be sanded to remove irregularities before any covering is put over it. If a thin underlayment or no underlayment is being used, wide joints between existing floorboards should be filled to avoid show-through on the less rigid types of finish floor. An underlayment of plywood or wood-based panel material installed over the old floor is required when linoleum or resilient tile is used for the new finish floor.

If underlayment is required, it should be in 4- by 4-ft or larger sheets of untempered hardboard, plywood, or particleboard. Some floor coverings are not guaranteed over all types of underlayments; check the manufacturer's recommendations before choosing an underlayment. Underlayment grade plywood has a sanded, C-plugged or better face ply and a C-ply or better veneer immediately under the face; interior or exterior grades as well as interior grades with exterior glue are available. The interior type is generally adequate for underlayment, but one of the other two types should be used if there is possible exposure to moisture. Underlayment should be laid with 1/32-in. spacing around all sides to permit expansion. Nail the underlayment to the subfloor by using the type of nail and the spacing recommended by the underlayment manufacturer.

Wood flooring, sheet vinyl with resilient backing, seamless flooring, and carpeting can be installed directly over the old flooring after major voids have been filled and after it has been sanded relatively smooth. These coverings can also be installed over old resilient tile that is still firmly cemented.

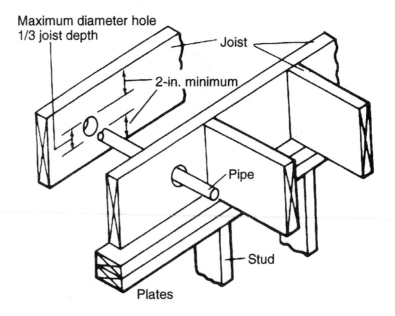

Figure 67—Drilled holes in joists.

Wood flooring may be hardwood or softwood. Grades and descriptions of wood flooring are listed in Table 8, and types are illustrated in Figure 68.

Hardwood flooring is available in strip or block patterns and is usually tongue-and-groove and end-matched, but it may be square-edged in thinner patterns. The most widely used pattern of hardwood strip flooring is 25/32 by 2-1/4 in. with hollow back. Strips are furnished in random lengths varying from 2 to 16 ft long. The face is slightly wider than the bottom so that tight joints result.

Table 8—Grades and descriptions of strip flooring of several species and ring orientation

Species	Grain orientation	Size (in.)		First grade	Second grade	Third grade
		Thickness	Face width			
		Softwoods				
Douglas-fir and hemlock	Edge grain	25/32	2-3/8–5-3/16	B and better	C	D
	Flat grain	25/32	2-3/8–5-3/16	C and better	D	– –
Southern Pine	Edge grain	5/16–	1-3/4 by 5-7/16	B and better	C and better	D (and no. 2)
	Flat grain	1-5/16				
		Hardwoods				
Oak	Quarter sawn	3/4	1-1/2–3-1/4	Clear	Select	No. 1 common
	Flat grain	3/8	1-1/2, 2			
		1-1/2, 2				
Beech, birch, maple, and pecan[a]		3/4	1-1/2 by 3-1/4	First grade	Second grade	Third grade
		3/8	1-1/2, 2			
		1/2	1-1/2, 2			

[a]Special grades are available in which uniformity of color is required.

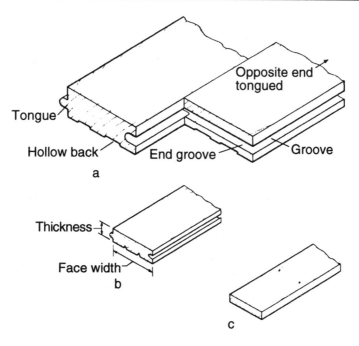

Figure 68—Strip flooring: (a) side- and end-matched, (b) side-matched, (c) square-edged.

Softwood flooring is also available in strip or block patterns. Strip flooring has tongue-and-groove edges, and some types are end-matched. Softwood flooring costs less than most hardwood species, but is less wear-resistant and shows surface abrasions more readily. It is best used in light traffic areas.

Bundles of flooring should be opened and kept in a heated space until the moisture content common to interior finish in the locale is achieved.

Strip flooring is normally laid crosswise to the floor joists; however, a new layer of strip flooring should be laid crosswise over old strip flooring. Nail sizes and types vary with the thickness of the flooring; use 8d flooring nails for 25/32-in. flooring, 6d flooring nails for 1/2-in. flooring, and 4d casing nails for 3/8-in. flooring. Other nails, such as the ring-shank and screw-shank types, can be used, but it is wise to check the flooring manufacturer's recommendations on size and diameter for specific uses. Flooring brads with blunted points, which prevent splitting of the tongue, are also available.

Begin installing matched flooring by placing the first strip 1/2 to 5/8 in. away from the wall to permit expansion when the moisture content increases. Nail straight down through the board near the grooved edge (Fig. 69). The nail should be close enough to the wall to be covered by the base or shoe molding and should be driven into a joist when the flooring is laid crosswise to the joists. The tongue should also be nailed, and consecutive flooring boards should be nailed through the tongue only. Nails are driven into the tongue at an angle of 45° to 50° and are not driven quite flush, to prevent damaging the edge with the hammer head (Fig. 70). The nail is then set with the end of a large nail set or by laying the nail set flatwise against the flooring. Contractors use devices designed especially for nailing flooring that drive and set the nail in one operation.

Choose the lengths of flooring boards so that butts will be well-separated in adjacent courses. Drive each board tightly against the one previously installed. Crooked boards should be forced into alignment or cut off and used at the ends of a course or in closets.

The last course of flooring should be stopped 1/2 to 5/8 in. from the wall, as recommended for starting the first course. Face-nail the final course near the edge where the base or shoe will cover the nail.

Square-edged strip flooring must be installed over a substantial subfloor and can only be face-nailed. The installation procedures relative to spacing at walls, spacing of joints, and general attachment are the same as those for matched flooring.

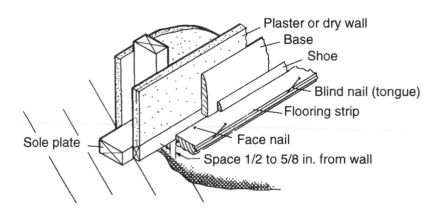

Figure 69—Installation of first strip of flooring.

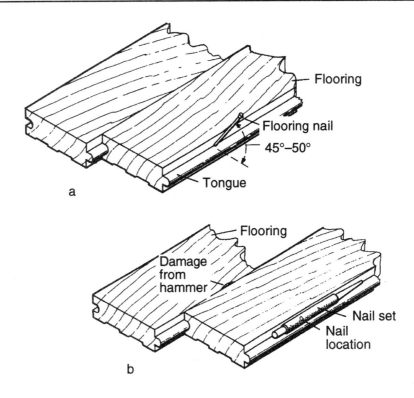

Figure 70—Nailing of flooring: (a) angle of nail, (b) setting the nail without damage to flooring.

Most wood or wood-based tile is applied with an adhesive to a smooth base, such as underlayment or a finished concrete floor with a properly installed vapor retarder. Do not apply wood to damp concrete floors; serious warping will result. Wood tile may be made up of a number of narrow slats held together by a membrane, cleats, or tape to form a square, or it may consist of plywood with tongue-and-groove edges (Fig. 71). To install wood tile, an adhesive is spread on the concrete slab or underlayment with a notched trowel and the tile is laid in the adhesive. Follow the manufacturer's recommendation for adhesive and its method of application. Wood-block flooring may have tongues on two edges and grooves on the other two edges, and it is usually nailed through the tongue into a wood subfloor. It may be set on concrete with an adhesive. The effects of shrinkage and swelling on wood block flooring are minimized by changing the grain direction of alternate blocks.

Particleboard tile is installed in much the same manner as wood tile, except it should not be used over concrete. Manufacturer's instructions for installation are usually quite complete. This tile is usually 9 by 9 by 3/8 in., with tongue-and-groove edges. The back is often marked with small saw kerfs to stabilize the tile and provide a better key for the adhesive.

Sheet vinyl with resilient backing smooths out minor surface imperfections. Most vinyl will lay flat with no adhesive required. Double-faced tape is used at joints and around the edge to keep the covering from moving. Most sheet vinyls are available in widths of 6, 9, 12, and 15 ft, so complete rooms can be covered with a minimum of splicing. This complete coverage permits fast, easy installation. The material is merely cut to room size with scissors, then taped down.

Seamless flooring consisting of resin chips combined with a urethane binder can be applied over any stable base, including old floor tile. This material is applied as a liquid in several coats and allowed to dry between coats. Complete application may take from 1/2 to 2 days, depending on the brand used. Manufacturer's instructions for installation should be followed. This

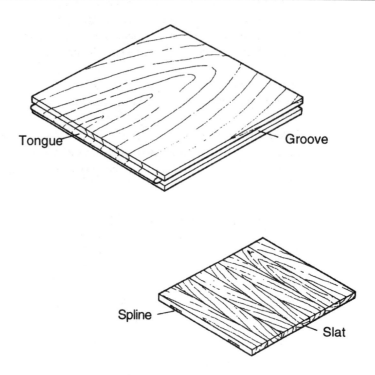

Figure 71—Two types of wood block flooring.

floor covering is easily renewed by additional coatings, and damaged spots are easily patched by adding more chips and binder.

Carpeting lends itself to rehabilitation and can be installed over almost any flooring that is level, relatively smooth, and free from major surface imperfections. Carpeting is available for all rooms in the house, including the kitchen. A very close weave is used for kitchen carpeting so that spills stay on the surface and are easily wiped up. Compared to wood, carpeting requires less maintenance, absorbs sound, and resists impact.

Linoleum and resilient tile both require a smooth underlayment or a smooth concrete slab to which they are bonded with adhesive. Linoleum should not be laid on concrete slabs on the ground floor or on basement floors, but many resilient tiles can be used in this way.

Linoleum is available in a variety of thicknesses and grades, and it is usually furnished in 6-ft-wide rolls. It is laid in accordance with manufacturer's directions and usually pressed to the floor with a roller to ensure adhesion.

One low-cost resilient covering is asphalt tile. Some types are damaged by grease and oil and therefore should not be used in kitchens. This tile is about 1/8 in. thick and has either 9- by 9-in. or 12- by 12-in. dimensions. Adhesive for this tile is spread with a notched trowel, and both the size of the notches and the adhesive are recommended by the manufacturer.

Other types of tile, such as vinyl, rubber, and cork, are also usually available in 9- by 9-in. or 12- by 12-in. size, but are sometimes larger. It is important that all these tiles be laid so the joints do not coincide with the joints of the underlayment. The manufacturer's directions usually include instructions for laying baselines near the center of the room and parallel to the length and width of the room. The baselines are then used as a starting point for laying the tile.

In the past, asbestos was added to some vinyl floor tiles and vinyl sheet flooring to provide greater strength. If this material is sanded, scraped, cut, or otherwise damaged, the asbestos fibers can be released. Appropriate safety precautions are discussed in detail in "Asbestos" under "Hazard Control" in the "Rehabilitation" section.

Wall Systems

Wall framing includes vertical studs and horizontal members, such as the top plate and bottom (or sole) plate and window and door headers (Fig. 72).

Exterior walls may be load-bearing (that is, supporting ceilings, upper floor, and/or roof), or they may be nonload-bearing (that is, not supporting a structural load, as under the gable end of a one-story house). Wall framing also serves as a nailing base for wall-covering materials and trim.

Framing

Wall-framing members are generally 2- by 4-in. studs spaced 16 or 24 in. on center, depending on vertical loads and the support requirements of the covering materials. Top plates and bottom plates are also 2 by 4 in. In some cases, if extra thick insulation is desired, 2- by 6-in. studs and plates have been used.

Headers over doors or windows in load-bearing walls consist of doubled 2- by 6-in. or wider members, depending on the span of the opening.

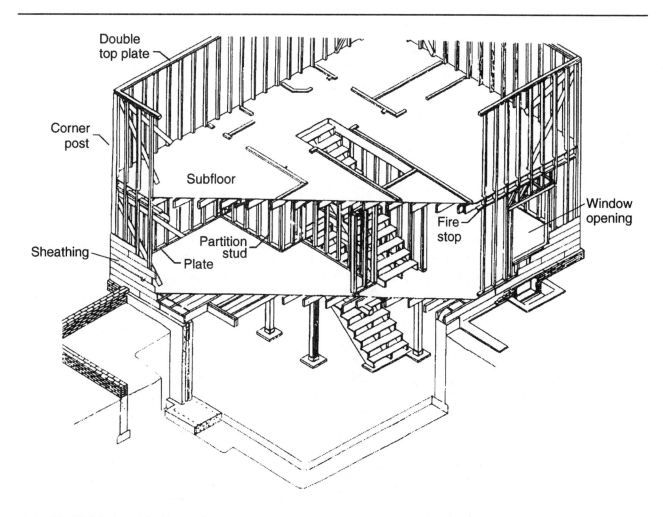

Figure 72—Wall framing for two-story house.

Figure 73 shows a commonly used method of wall framing for platform construction in 1-1/2- or 2-story houses with finished rooms above the first floor. Figure 74 shows wall framing for the second story.

The edge floor joist is toe-nailed to the top wall plate with 8d nails spaced 24 in. on center. The subfloor and wall framing are then installed in the same manner as the first floor.

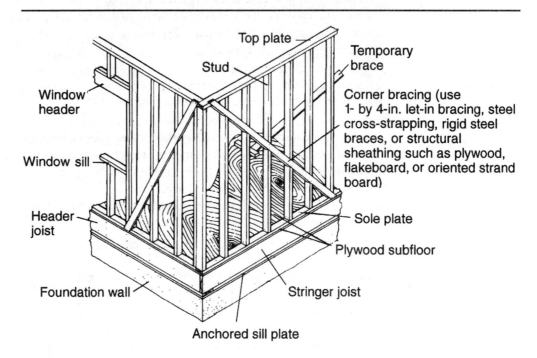

Figure 73—Wall framing with platform construction.

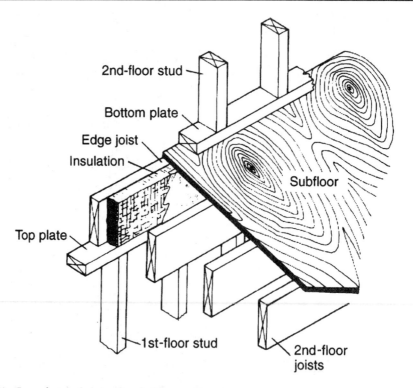

Figure 74—Second-story framing for platform construction.

The members used to span over window and door openings are called headers or lintels (Fig. 75). As the span of the opening increases, the depth of these members must be increased to support the ceiling and roof loads (see "Wall Openings" under "Major Structural Features" in the "Rehabilitation" section). A header is traditionally made up of two 2-in. members spaced with 1/2-in. lath or plywood strips, all of which are nailed together before installation for convenient handling. The lath or plywood spacers are used to bring the faces of the header flush with the edges of the studs. Light loads may require only a single header member. Headers are supported at the ends by the inner studs (jack studs), which are located on the inside of the window or door opening. Species and grades of wood normally used for floor joists are appropriate for headers.

Sheathing

Exterior wall sheathing is the covering applied over the framework of studs, plates, and window and door headers. It forms a base upon which the exterior siding can be applied.

In most older houses, 6-, 8-, and 10-in.-wide square-edged boards applied horizontally or diagonally were used as sheathing. Applying the boards diagonally eliminates the need for corner bracing. Board sheathing has been almost totally replaced by plywood, waferboard, oriented strandboard, or rigid insulation boards.

Sheathing should be added to the outside of the repaired parts of the house. Fifteen-pound asphalt paper is generally applied over the outside of the sheathing to help control air infiltration and to resist the entry of liquid water. This sheathing or breather paper should have a perm value of 60 or more to allow movement of water vapor from the inside to the outside. Vapor retarders should be placed on the warm side of exterior walls.

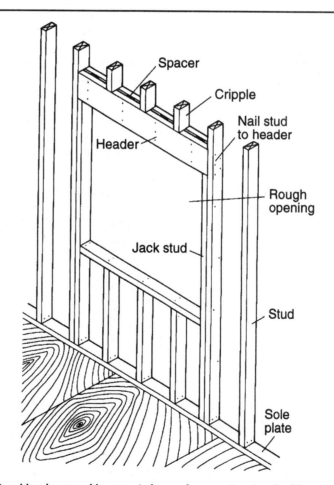

Figure 75—Traditional header assembly over window or door openings in a load-bearing wall.

Damaged wall systems are difficult to repair because the structural members are covered on the outside with the sheathing and finish siding material, and on the inside usually with plaster, drywall, or other panel material. Nonetheless, the structural members of the wall can suffer damage from decay or insects, and they can also tilt out of plumb. If the decay or insect damage is severe, the wall must be opened up enough to allow new framing members to be installed as reinforcements. Reinforcement is usually done by scabbing a shorter piece onto an existing member and nailing it firmly in place. The new piece should be about twice the length of the damaged area being replaced. Metal joint connectors, corner braces, and such, may help to fasten difficult-to-reach areas. Be careful not to damage any of the electrical wiring, plumbing, or other utilities that could be hidden in the wall.

If there is severe damage and various portions of the building have begun to sag or tilt out of plumb, heavy-duty jacks will be needed to level and square the building again. Once this is accomplished, new structural members should be firmly nailed or otherwise fastened in place. Diagonal bracing may also be needed to help keep the structure square. After the repairs are made, slowly release the pressure on the jacks and determine if the building will remain in the desired position. Fastening plumb lines at various locations and determining if the structure remains plumb over time may be helpful. Professional house movers may also be of assistance in straightening a badly deformed structure.

Wall Openings

Windows, exterior doors, and their frames are millwork items that are now fully assembled and delivered to the building site ready for installation. In the past, they were often custom made for a particular job by local millwork houses. Many millwork manufacturers are still capable of making these items to special order but, because of the setup time and hand labor that is involved, custom work is expensive.

Windows are available in many styles, including single- or double-hung, casement, stationary, awning, and horizontal sliding (Fig. 76). They can be made of wood, metal, vinyl, or a combination of wood or metal clad with vinyl. Window units may be purchased with either interior or exterior storm windows.

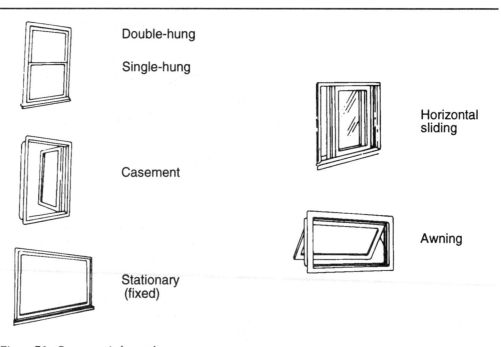

Double-hung

Single-hung

Casement

Stationary
(fixed)

Horizontal
sliding

Awning

Figure 76—Common window styles.

Glazing can consist of a single layer of glass or double- or triple-layer insulating glass. With insulating glass, the sheets are separated by a space that is evacuated and hermetically sealed. This type of glass offers better resistance to heat flow in or out of the house.

Wood window and doorframes should be treated with a water-repellent preservative for limited protection against dimensional change and decay.

Windows Windows may need repair, replacement, or relocation. If infiltration is not a serious problem and the wood is sound, repair should be considered. However, if the joints are decayed, replacement may be necessary. This is not difficult if the window size is not to be changed, unless that size is no longer made. Replacing the window will also mean reframing it or buying custom-made windows.

The sequence of window replacement will depend on the type of siding used. If new panel siding is being applied, the window is installed after the siding. If horizontal siding is used, the window must be installed before the siding.

If the wood in windows is showing signs of deterioration but the window is still in good operating condition, a water-repellent preservative may arrest further decay. First, remove existing paint, make sure that the wood is thoroughly dried, and brush on the preservative. Then let it dry and repaint the window. Paint cannot be used over some preservatives, so make sure to use a paintable water-repellent preservative.

A double-hung sash may bind against the stops, the jambs, or the parting strips. If this is the case, try waxing the parts in contact before doing any repair. If this does not eliminate the problem, try to determine where the sash is binding. Excessive paint buildup is a common cause of sticking and can be corrected by removing paint from stops and parting strips. A nailed stop (Fig. 77) can be moved away from the sash slightly, but if it is fastened by screws, it will probably be easier to remove the stop and hand-plane it lightly on the face contacting the

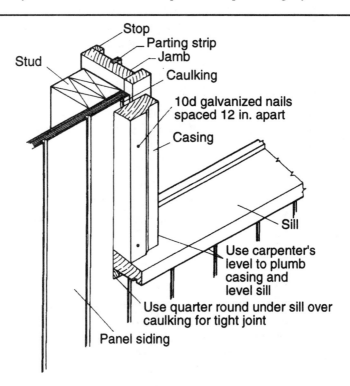

Figure 77—Installation of double-hung window frame.

125

sash. Loosening the contact between sash and stop too much will allow excessive air infiltration at the window. If the sash is binding against the jamb, remove the sash and plane the vertical edges slightly.

Adding full-width weatherstripping and spring balance units will provide a good airtight window that will not bind. These items are easily installed, requiring only removal of parting strip and stops. Install the units in accordance with the manufacturer's instructions and replace the stops.

If windows require extensive repairs, it will probably be more economical to replace them. New windows are usually purchased as a complete unit, including sash, frame, and exterior trim. These units are easily installed where a window of the same size and type is removed. Many older houses have tall, narrow windows that are no longer stock items, so in some cases it may be desirable to change the size or type of window. Most window manufacturers list rough-opening sizes for each of their windows. Here are some general rules to follow for rough-opening size:

1. Double-hung window (single unit):
    Rough opening width = glass width plus 6 in.
    Rough opening height = total glass height plus 10 in.

2. Casement window (two sash):
    Rough opening width = total glass width plus 11-1/4 in.
    Rough opening height = total glass height plus 6-3/8 in.

After the old window is removed, take off the interior wall covering to the rough-opening width for the new window. If a larger window must be centered in the same location as the old one, half the necessary additional width may be cut from each side; otherwise, the entire additional width may be cut from one side. For windows 3-1/2 ft or less in width, no temporary support of ceiling and roof should be required. If windows more than 3-1/2 ft wide are to be installed, provide some temporary support for the ceiling and roof before removing existing framing in bearing walls. Remove framing to the width of the new window, and frame the window as shown in Figure 78.

The header must be supported at both ends by cripple studs (jack studs). Headers are made up of two 2-in.-thick members, usually spaced with lath or plywood strips to produce the same thickness as the nominal 2- by 4-in. stud space. The following sizes might be used as a guide for headers:

Maximum span (ft)	Header size (nominal in.)
3	Two 2 by 6
5	Two 2 by 8
6	Two 2 by 10
8	Two 2 by 12

Independent design may be necessary for wider openings.

Cut the sheathing, or panel siding used without sheathing, to the size of the rough opening. If bevel siding is already in place, cut it to the size of the window trim so that the sheathing will butt against the window casing (Fig. 79). Determine where to cut the siding by inserting the preassembled window frame in the rough opening and marking the siding around the outside edge of the casing.

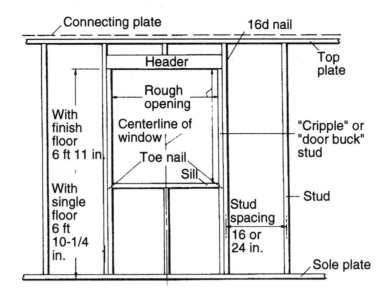

Figure 78—Framing at window opening and height of window and door headers.

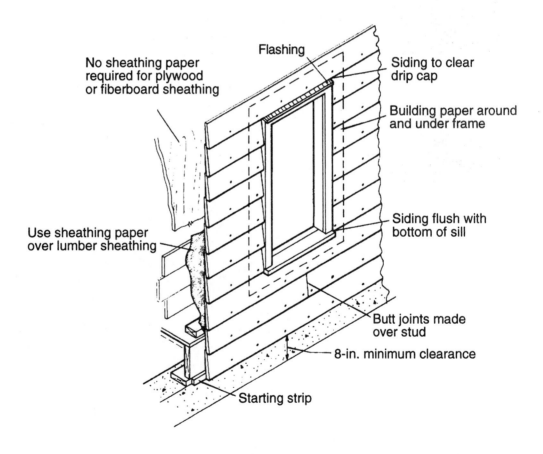

Figure 79—Application of bevel siding to coincide with window sill and drip cap.

Before installing the window frame in the rough opening, take precautions to ensure that water and wind do not come in around the finished window. If panel siding is used, place a ribbon of caulking sealant (rubber or similar base) over the siding at the location of the side and head casing (Fig. 80). If horizontal siding is used over sheathing, loosen the siding around the opening and slide strips of 15-lb asphalt felt between the sheathing and siding around the opening (Fig. 79).

Place the frame in the rough opening, preferably with the window closed to keep it square, and level the sill with a carpenter's level. Use shims under the sill on the inside if necessary. The shims should be positioned under each point where nails will be driven so that the nails do not cause the window casing to bend. Plumb the side casing and jamb for level and squareness; then nail the frame in place with 10d galvanized nails. Nail through the casing into the side studs and header (Fig. 77), spacing the nails about 12 in. apart. When double-hung windows are used, slide the sash up and down while nailing the frame to be sure that it works freely. To install the frame over panel siding, place a ribbon of caulking sealant at the junction of the siding and the sill, and install a small molding, such as quarter round, over the caulking. Moving a window to a different location, or eliminating a window and adding a new one at another location, may be desirable. If a window is removed, close the opening by adding 2- by 4-in. vertical framing members spaced no more than 16 in. apart. Keep framing in line with existing studs under the window or in sequence with wall studs so covering materials can be nailed to them easily. Toe-nail new framing to the old window header and to the sill using three 8d or 10d nails at each joint. Install sheathing of the same thickness as the original, add insulation, and apply a vapor retarder on the inside face of the framing. Make sure that the vapor retarder covers the rough framing of the existing window and that it overlaps any vapor retarder in the remainder of the wall. (Insulation and vapor retarders are discussed in detail in other sections.) Apply interior and exterior wall covering to match the existing coverings on the house.

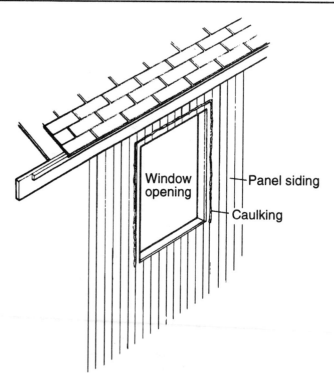

Figure 80—Caulking around window opening before installing frame.

In cold climates, storm windows are necessary for comfort, for economical heating, and to avoid damage from excessive condensation on the inside face of the window. If old windows are not standard sizes, storm windows must be made by building a frame to fit the existing window and fitting glass to the frame. Storm windows are commercially available to fit all standard-size windows. One of the most practical types is the self-storing or combination storm-and-screen window. These need minor adjustments for width and height and can be custom-fabricated for odd-sized windows at a moderate cost.

Doors

Doors in houses of all ages frequently stick or fail to latch. To remedy the sticking door, first determine where it is sticking. If the frame is not critically out of square, some minor adjustments may remedy the situation. The top of the door could be planed or sanded without removing the door. If the side of the door is sticking near the top or bottom, the excess width can also be planed off without removing the door; however, the edge will have to be refinished or repainted. If the side of the door sticks near the latch or over the entire height of the door, remove the hinges and plane the hinge edge; additional routing is required before the hinges are replaced. If the door is binding on the hinge edge, the hinges may have been routed too deeply. This can be corrected by loosening the hinge leaf and adding a filler under it to bring it out slightly. If the latch does not catch, remove the strike plate and shim it out slightly. Replace the strike plate by first placing a filler, such as a matchstick, in the screw hole and reinserting the screw so that the strike plate is relocated slightly away from the stop. If exterior doors are badly weathered, it may be better to replace them rather than attempting to repair them. Doors can be purchased separately or with frames, including exterior side and head casing with rabbeted jamb and a sill (Fig. 81). Exterior doors should be either panel or solid-core flush type. Several styles are available, most of them featuring some type of glazing (Fig. 82).

Hollow-core flush doors should be limited to interior use except in warm climates because they warp excessively during the heating season when used as exterior doors. The standard height for exterior doors is 6 ft 8 in.; standard thickness is 1-3/4 in. The main door should be 3 ft wide, and the service or rear door should be at least 2 ft 6 in. wide, preferably up to 3 ft wide.

If rough framing is required either for a new door location or because the old framing is not square, provide header and cripple (jack) studs as shown in Figure 78. The rough opening height should be the height of the door plus 2-1/4 in. above the finished floor; width should be the width of the door plus 2-1/2 in. Use doubled 2- by 6-in. studs for headers, and fasten them in place with two 16d nails through the stud into each member. If the stud space on each side of the door is not accessible, toe-nail the header to the studs. Nail cripple studs, supporting the header on each side of the opening, to the full stud with 12d nails spaced about 16 in. apart and staggered. After sheathing or panel siding is placed over the framing, leaving only the rough opening, the doorframe can be installed. Apply a ribbon of caulking sealant on each side and above the opening where the casing will fit over it. Place the doorframe in the opening and secure it by nailing through the side and head casing. Nail the hinge side first. In a new installation, the floor joists and header must be trimmed to receive the sill before the frame can be installed (Fig. 83). The top of the sill should be the same height as the finish floor so that the threshold can be installed over the joint. Shim the sill when necessary so that it will have full bearing on the floor framing. If joists are parallel to the sill, headers and a short support member are necessary at the edge of the sill. Use quarter-round molding in combination with caulking under the door sill when a panel siding or other single exterior covering is used. Install the threshold over the junction with the finish floor by nailing it to the floor and sill with finishing nails.

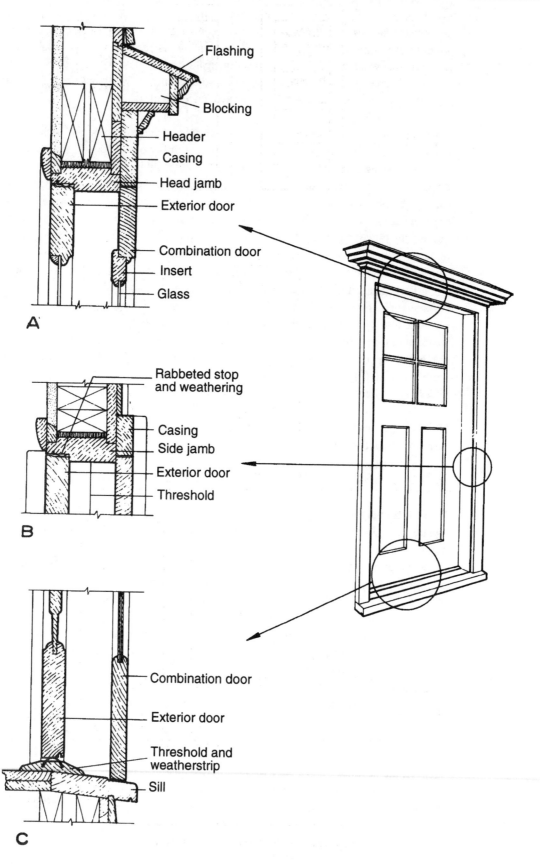

Figure 81—Exterior door and frame. Exterior door and combination door (screen and storm) cross sections: (A) head jamb, (B) side jamb, (C) sill.

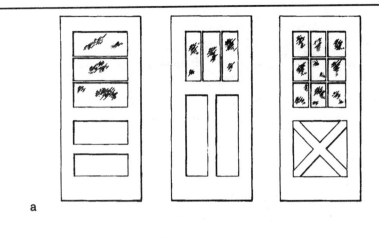

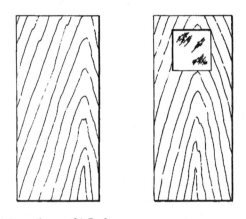

Figure 82—Exterior doors: (a) panel type, (b) flush type.

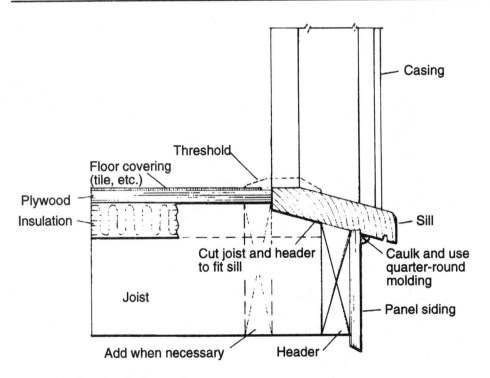

Figure 83—Door installation at sill.

Any trimming to reduce the width of the door is done on the hinge edge. Hinges are routed or mortised into the edge of the door with about 3/16- or 1/4-in. back spacing (Fig. 84). In-swinging exterior doors require 3-1/2- by 3-1/2-in. loose-pin hinges. Nonremovable pins are used on out-swinging doors. Use three hinges to minimize warping. Bevel edges slightly toward the side that will fit against stops. Clearances are shown in Figure 85. Carefully measure the opening width and plane the edge for the proper side clearances.

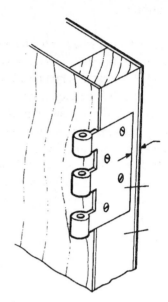

Figure 84—Installation of door hinges.

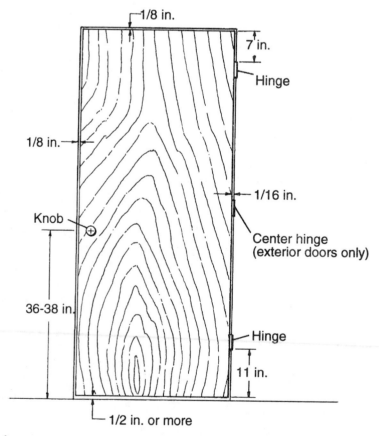

Figure 85—Door clearances.

Next, square and trim the top of the door for proper fit; then saw off the bottom for the proper floor clearance. All edges should then be sealed to minimize entrance of moisture. Exterior doors are usually purchased with an entry lock set that is easy to install.

In cold climates, weatherstrip all exterior doors. Check weatherstripping on old doors, and replace it if it shows wear. Also consider adding storm doors in cold climates.
Storm doors not only save heat but protect the surface of the exterior door from the weather and help prevent warping.

If a new interior door is added or the framing is replaced, the opening should be rough-framed in a manner similar to that for exterior doors. The rough-framing width is the door width plus 2-1/2 in.; height is the door height plus 2 in. above the finished floor. The head door jamb and two side door jambs are the same width as the overall wall thickness if wood casing is used. If metal casing is used with drywall (Fig. 86), the jamb width is the same as the stud depth. Jambs are often purchased in precut sets and can even be purchased complete with stops and with the door prehung in the frame. Jambs also can be made in a small shop with a table or radial-arm saw (Fig. 87). The prehung door is by far the simplest to install and is usually the most economical because of the labor savings. Even if the door and jambs are purchased separately, the installation is simplified by prehanging the door in the frame at the building site. The door then serves as a jig to set the frame in place.

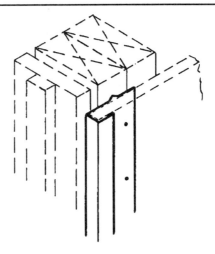

Figure 86—Metal casing used with drywall.

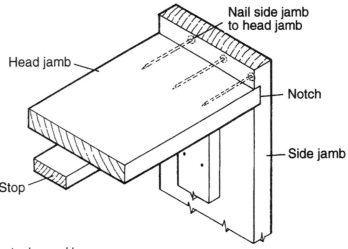

Figure 87—Door jamb assembly.

Before installing the door, temporarily put in place the narrow wood strips used as stops. Stops are usually 7/16 in. thick and may be 1-1/2 to 2-1/4 in. wide. Install them with a miter joint at the junction of the head and side jambs. A 45° bevel cut at the bottom of the stop, 1 to 1-1/2 in. above the finish floor, will eliminate a dirt pocket and make cleaning easier (Fig. 88). This is called a sanitary stop.

Fit the door to the frame, using the clearances shown in Figure 85. Bevel the edges slightly toward the side that will fit against the stops. Rout or mortise the hinges into the edge of the door with about a 3/16- or 1/4-in. back spacing. Make adjustments, if necessary, to provide sufficient edge distance so that the screws penetrate the wood. For interior doors, use two 3- by 3-in. loose-pin hinges. If a router is not available, mark the hinge outline and depth of cut and remove the wood with a wood chisel. The surface of the hinge should be flush with the wood surface. After attaching the hinge to the door with screws, place the door in the opening, block it for proper clearances, and mark the location of door hinges on the jamb. Remove the door and rout the jamb to the thickness of the hinge half. Install the hinge halves on the jamb, place the door in the opening, and insert the pins.

Lock sets are classed as (a) entry lock sets (decorative keyed lock), (b) privacy lock set (inside lock control with a safety slot for opening from the outside), (c) lock set (keyed lock), and (d) latch set (without lock). The lock set is usually purchased with the door and may even be installed with the door. If the lock set is not installed, directions are provided, including paper templates that facilitate exact location of holes. After the latch is installed, mark its location on the jamb when the door is in a near-closed position. Mark the outline of the strike plate for this position and rout the jamb so that the strike plate will be flush with the face of the jamb (Fig. 89).

The stops, which were temporarily nailed in place, can now be permanently installed. Nail the stop on the lock side first, setting it against the door face when the door is latched. Nail the stops with finishing nails or brads 1-1/2 in. long and spaced in pairs about 16 in. apart. The stop at the hinge side of the door should allow a clearance of 1/32 in. (Fig. 90).

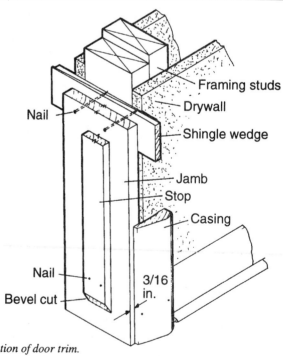

Figure 88—Installation of door trim.

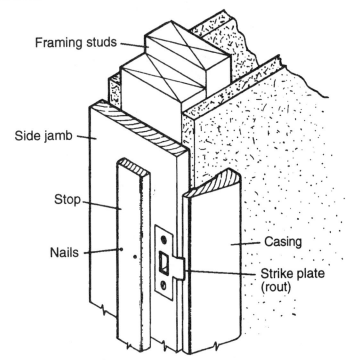

Figure 89—*Installing door strike plate.*

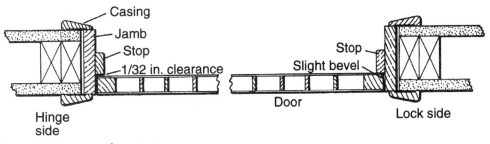

Figure 90—*Door stop clearances.*

To install a new doorframe, place the frame in the opening and plumb and fasten the hinge side of the frame first. Use shingle wedges between the side jamb and the rough door buck to plumb the jamb (Fig. 88). Place wedge sets at the hinge and latch locations plus intermediate locations along the height, and nail the jamb with pairs of 8d nails at each wedge. Continue the installation by fastening the opposite jamb in the same manner. After the door jambs are installed, cut off the shingle wedges flush with the wall.

Casing is the trim around the door opening. Shapes are available in thicknesses from 1/2 to 3/4 in. and widths from 2-1/4 to 3-1/2 in. A number of styles are available (Fig. 91). Metal casing used at the edge of the drywall eliminates the need for wood casing (Fig. 86).

Position the casing with about a 3/16-in.-edge distance from the face of the jamb (Fig. 88). Nail it in place with 6d or 7d casing or finishing nails, depending on the thickness of the casing. Casings with one thin edge should be nailed with 1-1/2-in. brads along the edge. Space nails in pairs about 16 in. apart. Casings with molded forms must have a miter joint where the head and side casings join (Fig. 92a), but rectangular casings are butt-jointed (Fig. 92b).

Metal casing can be installed by either of two methods. In one method, the casing is nailed to the door buck (door framing studs) around the opening; then the drywall is inserted into the groove and nailed to the studs in the usual fashion. The other method consists of fitting the casing over the edge of the drywall, positioning the sheet properly, and then nailing through the

135

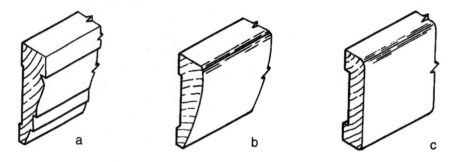

Figure 91—Styles of door casings: (a) colonial, (b) ranch, (c) plain.

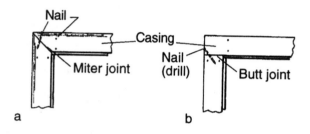

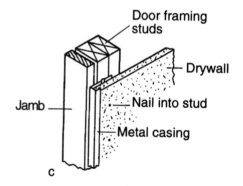

Figure 92—Installing door trim: (a) molded casing, (b) rectangular casing, (c) metal casing.

drywall and casing into the stud behind it (Fig. 92c). Use the same type of nails and spacing as for drywall alone.

Interior doors are either panel or flush style. Flush doors (Fig. 93) are usually hollow core. Moldings are sometimes included on one or both faces. Such moldings can also be applied to existing doors if added decoration is desired. Panel-type doors are available in a variety of patterns. Two popular patterns are the five-cross-panel and the colonial.

Standard door height is 6 ft 8 in.; however, a height of 6 ft 6 in. is sometimes used with low ceilings, such as in the upstairs of a story-and-a-half house or in a basement. Door widths vary, depending on use and personal taste; however, minimums may be governed by building regulations. Usual widths are bedroom and other rooms, 2 ft 6 in.; bathroom, 2 ft 4 in.; and

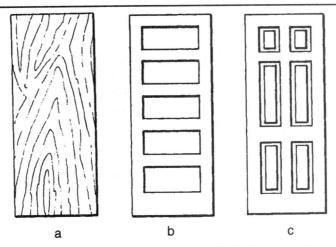

Figure 93—Interior doors: (a) flush, (b) five-cross-panel, (c) colonial panel type.

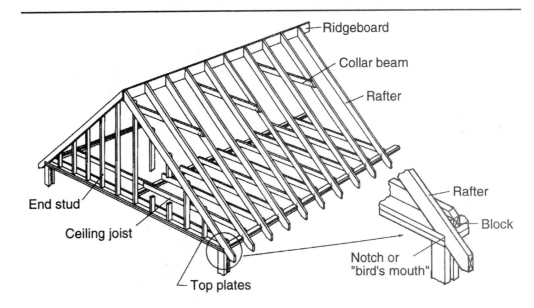

Figure 94—Typical rafter framing for pitched roof.

small closet and linen closet, 2 ft. If the size of a hollow-core door is changed significantly, the framing within the door will need to be replaced.

Roof Systems

Construction Techniques

Roof systems have traditionally been composed of rafters, sheathing, ridge boards, and collar beams (Fig. 94); roof trusses are more typical of current practice. Roof frames provide structural members to which roofing, vents, and finish ceiling materials may be attached and within which insulation materials may be placed. Pitched roofs create room for storage and living space that costs less than main floor space because no additional foundation is required and roof costs do not increase proportionately with the increase in living area.

The more common styles of roofs include gable, gable with dormers, and hip roofs (Fig. 95). Mansard, gambrel, A-frame, and flat roofs are also popular. The type of roof greatly affects the appearance of the house.

Roofs are particularly critical because they protect the rest of the house from rain and weather. Once the roof starts to fail (leak), the rest of the house will be damaged substantially in a very short time. Therefore, it is imperative to correct any deficiency promptly. Roof overhang is

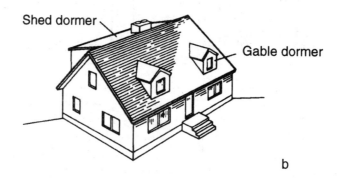

Figure 95—Pitched roof types: (a) gable, (b) gable with dormers, (c) hip.

important because it protects the side walls from rain and provides some shade. If properly sized, overhangs allow sunlight to penetrate south-facing windows in the winter when solar heat is desired, but not in the summer when the sun is at a higher angle in the sky.

The slope of a roof is generally expressed as the number of inches of vertical rise in 12 in. of horizontal run. The rise is given first; for example, 4-in-12 or 4/12 pitch. Older homes often have relatively steep roofs, making repair work difficult and dangerous.

Rehabilitation of Damaged Systems

Some roofing shingles were manufactured with asbestos. Breaking or removing these shingles can release asbestos fibers, so appropriate safety precautions should be taken. The handling of asbestos-containing materials is discussed in detail in "Asbestos" under "Hazard Control" in the "Rehabilitation" section.

Rafters transfer the load they are carrying to their bottom end and tend to push the side walls out. As a result, rafters tend to sag at the ridge line. Rafters are restrained from sagging by the ceiling joists that are placed in tension and consequently must be securely fastened to the rafters and to each other where spliced. The use of collar beams (Fig. 94) also helps to keep rafters from sagging.

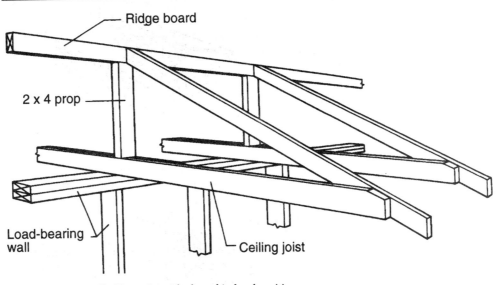

Ridge board

2 x 4 prop

Load-bearing wall

Ceiling joist

Figure 96—Prop to hold sagging ridgeboard in level position.

A sagging ridgeboard can sometimes be leveled by jacking it at points between supports and installing props to hold it in a level position. When jacking the ridgeboard, the jack must be located where the load can be traced down through the structure so that the ultimate bearing is directly on the foundation. If there is no conveniently located bearing partition, install a beam under the ridge and transfer the load to bearing points. After the ridgeboard is jacked to a level position, cut a 2- by 4-in. stud just long enough to fit between the ceiling joist and ridgeboard or beam and nail it at both ends (Fig. 96). For a short ridgeboard, one prop may be sufficient.

Add props as needed. In some repairs, it may be sufficient to add collar beams without props.

If rafters are sagging, nail a new rafter to the side of the old one after forcing the new rafter ends into position. Permanent set in the old rafters cannot be removed.

The roof sheathing may have sagged between rafters, resulting in a wavy roof surface. If this condition exists, new sheathing is required. Often the sheathing can be nailed directly over the old roofing. This saves the labor of removing old roofing and the associated cleanup. Wood shingles that show any indication of decay should be completely removed before new sheathing is applied. If wood shingles are excessively cupped or otherwise warped, they should also be removed. Wood-shingle and slate roofs on older houses were often installed on furring strips rather than on solid sheathing.

Sheathing nailed over existing sheathing or over sheathing and roofing must be secured with longer nails than would normally be used. Nails should penetrate the framing 1-1/4 to 1-1/2 in. Nail edges of all sheathing panels at 6-in. spacing and to intermediate framing members at 12-in. spacing. Apply the panels with the long dimension perpendicular to the rafters. If the sheathing does not have tongue-and-groove edges, use clips at unsupported edges on a built-up roof. Clips are commercially available and should be installed in accordance with the fabricator's instructions.

For 16-in. rafter spacing, 3/8-in. plywood is the minimum thickness that should be used, although 1/2-in.-thick plywood is preferable. Structural flakeboard panels are comparable with plywood in strength and water resistance, which are key properties for exterior sheathing. Rafter or truss spacing, nailing, and edge treatments are the same as those for plywood of the same thicknesses. Some flakeboard panels, called oriented strandboard, have the individual wood particles aligned to increase the directional strength parallel to the length of the panel. These products should be laid with the longer dimension perpendicular to the supports.

Roofs that have been allowed to deteriorate excessively are dangerous. Wornout shingles allow moisture to contact the sheathing and rafters, which causes decay. Wood members that are even slightly decayed often have substantially reduced strength and could fail without warning. Obviously, they must be replaced.

Addition of Roof Overhang

The addition of a roof overhang will soon pay for itself in reduced maintenance on siding and exterior trim. Without the overhang, water washes down the face of the wall, creating moisture problems in the siding and trim, and consequently, more frequent painting is required. Roof overhang also improves the appearance of the house and properly shades the windows.

When new sheathing is being added, it can be extended beyond the edge of the existing roof to provide some overhang. This is a minimum solution, and the extension should not be more than 12 in. if 1/2-in. plywood sheathing is used. Any greater extension would require some type of supporting framing. Framing can usually be extended at the eave by adding to each rafter. First, remove the frieze board, or in the case of a closed cornice, remove the fascia. Nail a 2 by 4 to the side of each rafter, letting it extend beyond the wall to the amount of the desired overhang. The 2 by 4 should extend inside the wall a distance equal to the overhang (Fig. 97a).

Framing for an overhang at the gable ends can be made by adding a box frame. The ridge beam and eave facia must be extended to support this boxed framing. An alternate extension is possible by placing a 2 by 4 flat, cutting into gable framing, and extending the 2 by 4 back to the first rafter (Fig. 97b).

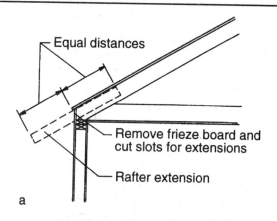

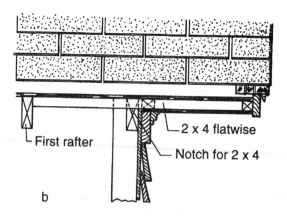

Figure 97—Extending roof overhang: (a) rafter extension at eaves, (b) extension at gable end.

Roof Coverings   A wide variety of roof coverings is available, and most can be used for rehabilitation in the same manner as for new construction. Sometimes there are local code requirements for fire safety. Cost usually influences the choice. In most houses, the roof is a major design element, so the covering material must fit the house design. Heavy materials such as tile or slate should not be used unless they replace the same material or unless the roof framing is strengthened to support the additional load. The most popular covering materials for pitched roofs are wood or asphalt shingles. These can be applied directly over old shingles or over sheathing as described later in this section; however, if two layers of shingles exist from previous reroofing, it may be preferable to remove the old roofing. Roll roofing is sometimes used for particularly low-cost applications or over porches with relatively low-pitched roofs. The most common covering for flat or low-pitched roofs is built-up roofing with a gravel topping.

If shingles will be applied over old wood or asphalt shingles, the industry recommends certain preparations. First, remove about 6-in.-wide strips of old shingles along the eaves and gables, and apply nominal 1-in. boards at these locations. Thinner boards may be necessary if application is over old asphalt shingles. Remove the shingles from ridges or hips and replace with bevel siding with the butt edges up. Fill each valley with lumber to separate the old metal flashing from the new. Double the first shingle course.

An underlay of 15- or 30-lb asphalt-saturated felt or roll roofing should be used in moderate- and lower-slope roofs covered with asphalt shingles, slate shingles, or tile roofing. Roll roofing 36 in. wide is also required at all valleys. Underlayment is not commonly used under wood shingles or shakes. A 45-lb or heavier smooth-surface roll roofing should be used as a flashing along the eave line in areas where moderate to severe snowfalls occur. The flashing should extend up the roof to a point 36 in. inside the warm wall. If two strips are required, use mastic to seal the joint. Also use mastic to seal end joints. This flashing gives protection from ice dams (Fig. 98).

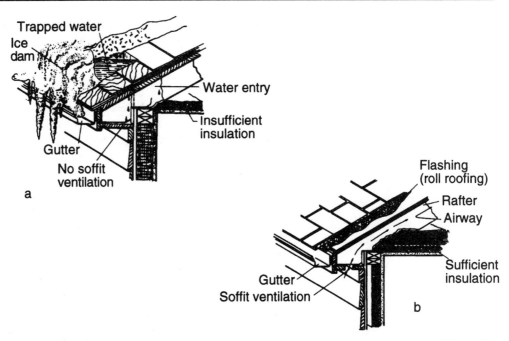

Figure 98—Flashing and ventilation to prevent damage from ice dams: (a) without proper flashing and ventilation, (b) proper flashing and ventilation.

Ice dams form when water from melting snow runs down the roof and freezes at the colder overhang. The ice gradually forms a dam that backs up water under the shingles. The wide flashing at the eave will minimize the chances of this water entering the ceiling or the wall. Good attic ventilation and sufficient ceiling insulation are also important in eliminating ice dams. This aspect of roof construction is described in "Attic" in the "Condition Assessment" section.

Wood shingles used for house roofs should be no. 1 grade, which are all heartwood, edge grain, and tapered. Principal species used commercially are western redcedar and redwood, which have heartwood with high resistance to decay and low shrinkage. Widths of shingles vary, and the narrower shingles are most often found in the lower grades. Recommended exposures for common shingle sizes are shown in Table 9.

When wood shingles or shakes are used in damp climates, it is common to space the roof boards. Wood nailing strips of nominal 1- by 3-in. or 1- by 4-in. size are spaced the same distance on centers as the shingle exposure. For example, if the shingle exposure to the weather is 5 in. and nominal 1- by 4-in. strips are used, there would be a space of 1-3/8 to 1-1/2 in. between adjacent boards.

These are general rules for applying wood shingles (Fig. 99):

1. Extend shingles 1-1/2 to 2 in. beyond the eave line and about 3/4 in. beyond the rake (gable) edge.

2. Nail each shingle with two rust-resistant (galvanized, stainless steel, or aluminum) nails spaced about 3/4 in. from the edge and 1-1/2 in. above the butt line of the next course. Use 3d nails for 16- and 18-in. shingles and 4d nails for 24-in. shingles. If shingles are applied over old wood shingles, use longer nails to penetrate through the old roofing and into the sheathing. A ring-shank nail (threaded) is recommended if the plywood roof sheathing is less than 1/2 in. thick.

3. Allow a 1/8- to 1/4-in. space between adjacent shingles in a course for expansion when wet. Lap vertical joints at least 1-1/2 in. by the shingles in the course above. Space the joints in succeeding courses so that the joint in one course is not in line with the joint in the second course above it.

4. Shingle away from valleys, selecting and precutting wide valley shingles. The valley should be 4 in. wide at the top and increase in width at the rate of 1/8 in. per foot from the top. Use valley flashing with a standing seam. Do not nail through the metal. Valley flashing should be a minimum of 24 in. wide for roof slopes under 4 in 12, 18 in. wide for roof slopes of 4 in 12 to 7 in 12, and 12 in. wide for roof slopes of 7 in 12 and flatter.

5. Place a metal edging along the gable end of the roof to aid in guiding the water away from the end walls.

Table 9—Recommended exposure for wood shingles

Shingle length (in.)	Shingle thickness (green)	Maximum exposure (in.)[a]	
		Slope less than 4 in 12	Slope 4 in 12 and steeper
16	Five butts in 2 in.	3-3/4	5
18	Five butts in 2-1/4 in.	4-1/4	5-1/2
24	Four butts in 2 in.	5-3/4	7-1/2

[a]Minimum slope for main roofs is 4 in 12; minimum slope for porch roofs is 3 in 12.

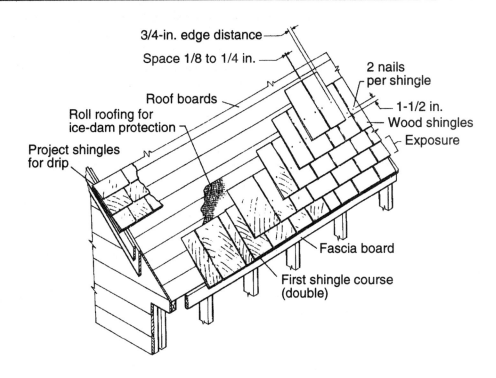

Figure 99—Application of wood-shingle roofing over boards.

Roll roofing felt is not required for wood shingles except for protection in ice dam areas.

Wood shakes are applied in much the same manner as shingles, except that longer nails must be used because the shakes are thicker. Shakes are exposed more than shingles because of their length. Exposures are 8 in. for 18-in. shakes, 10 in. for 24-in. shakes, and 13 in. for 32-in. shakes. Butts are often laid unevenly to create a rustic appearance. An 18-in.-wide underlay of 30-lb roll roofing should be used under each course to prevent wind-driven snow and rain from entering between the rough faces of the shakes. Position the bottom of the roll roofing above the butt edge of the shakes a distance equal to double the weather exposure. If exposure is less than one-third the total length, roll roofing is not usually required.

Asphalt shingles consist of a felt base with asphalt coating. Concealed spots of adhesive called seal tabs serve to glue the shingles together after they are installed and thus provide greater wind resistance. Felt-based, square-butt asphalt strip shingles should have a minimum weight of 240 lb per "square." Three bundles of 27 shingles each make a square and are enough to cover 100 ft² of roof surface.

Fiberglass shingles are made of an inorganic fiberglass base that is coated with asphalt. The usual minimum recommended weight is 220 lb per square. Fiberglass shingles are generally considered more durable and fire-resistant than felt-based shingles. Application methods are the same for both types.

The most common type of asphalt or fiberglass shingle is the square-butt strip shingle, which is 12 by 36 in., has three tabs, and is usually laid with 5 in. exposed to the weather. Bundles should be piled flat so that the strips will not curl when the bundles are opened for use. An underlayment of 15-lb roll roofing is often used between the sheathing and the shingles. Table 10 shows the requirements for applying underlayment.

Table 10—Underlayment requirements for asphalt shingles

Minimum roof slope[a]		Underlayment[b]
Double-coverage shingles	Triple-coverage shingles	
7 in 12	4 in 12[c]	Not required
4 in 12[c]	3 in 12[d]	Single
2 in 12	2 in 12	Double

[a]Double coverage for a 12- by 36-in. shingle is usually an exposure of about 5 in. and about 4 in. for triple coverage.
[b]Headlap for single coverage of underlayment should be 2 in. and for double coverage, 19 in.
[c]May be 3 in 12 for porch roofs.
[d]May be 2 in 12 for porch roofs.

Begin installing the roofing by first applying a metal edging along the gable end eave line (Fig. 100b). The first course of asphalt shingles is doubled and extended downward beyond the edging about 1/2 in. to prevent the water from backing up under the shingles. A 1/2-in. projection should also be used at the rake (Fig. 100b). Make several chalklines on the underlayment parallel to the roof slope to serve as guides in aligning the shingles so that tabs are in a straight line. Use manufacturer's directions in securing the shingles. Using seal-tab or lock shingles and nailing each 12- by 36-in. strip with six 1-in. galvanized roofing nails (Fig. 100) is a good practice in areas of high winds. If a nail penetrates a crack or knothole, remove the nail, seal the hole, and replace the nail in sound wood. If the nail is not in sound wood, it will gradually work out and cause a hump in the shingle above it.

Built-up roof coverings are limited to flat or low-pitched roofs and are installed by contractors who specialize in this work. The roof consists of three, four, or five layers of roofers' felt, with each layer mopped down with tar or asphalt. The final surface is then coated with asphalt and usually covered with gravel embedded in asphalt or tar.

Other roof coverings, such as slate, tile, and metal, should be installed by specialized applicators; their installation is not described in detail. These roof coverings are generally more expensive and are not used as widely as wood or asphalt shingles and built-up roofs.

The Boston ridge is the most common method of treating the roof ridge, and is also applicable to hip roofs. If asphalt shingles are used, cut the 12- by 36-in. strips into 12- by 12-in. sections. Bend them slightly and use in a lap fashion over the ridge with a 5-in. exposure distance (Fig. 101). The laps are turned away from the prevailing wind. In areas where there are driving rains, the use of metal flashing under the shingle ridge is recommended. The use of a ribbon of asphalt roofing cement under each lap will also reduce the chance of water penetration. Locate nails where they will be covered by the lap of the next section. A small spot of asphalt cement under each exposed edge will give a positive seal.

Wood-shingle roofs can also be finished with a Boston ridge. Flashing should be placed first over the ridge. Six-in.-wide shingles are alternately lapped, fitted, and blind-nailed (Fig. 102). Exposed shingle edges are alternately lapped.

A metal ridge roll can also be used on asphalt-shingle or wood-shingle roofs (Fig. 103). This ridge of copper, galvanized iron, or aluminum is formed to the roof slope.

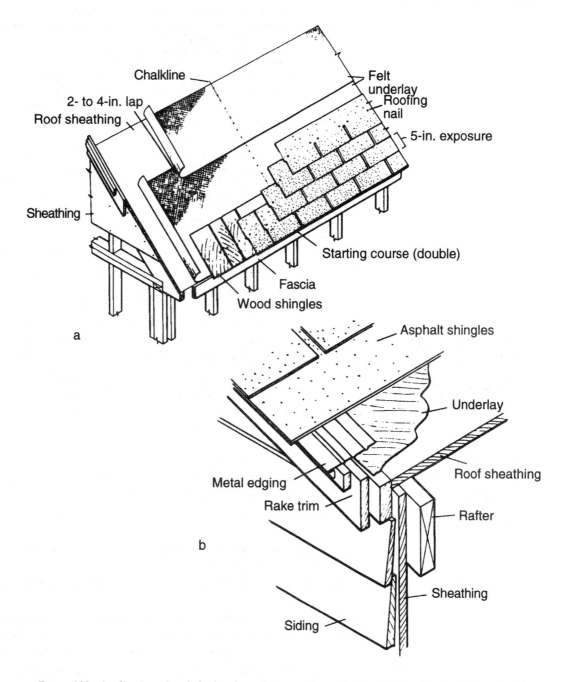

Figure 100—Application of asphalt-shingle roofing over plywood: (a) with strip shingles, (b) metal edging at gable end.

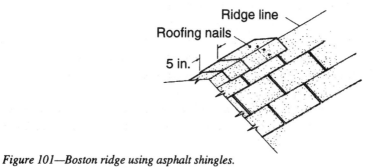

Figure 101—Boston ridge using asphalt shingles.

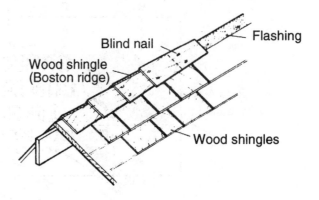

Figure 102—Boston ridge using wood shingles.

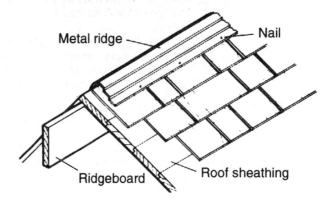

Figure 103—Metal ridge roll.

Repair of Structural
Decay and Insect and
Rodent Damage

*Decay From
Moisture*

The use of proper design, construction, and maintenance techniques is the best way to eliminate excess moisture and to prevent decay. If decayed wood is present in the structure, the source of moisture that caused the decay must be identified. If the source of moisture can be eliminated and serious decay has not resulted, no further action is needed. If there is serious decay, the wood members should be replaced, particularly if they serve a structural function. Because decay often progresses several feet beyond where it is visible to the naked eye, it may be necessary to replace more material than is obvious at first evaluation.

If the source of moisture cannot be eliminated, the decayed portions should be replaced with wood that has been pressure-treated with an appropriate preservative. Wood used in residential construction is most commonly treated with specially formulated waterborne chemicals and is available at local lumber yards. It generally has a greenish or brownish cast.

When selecting the material, be sure there is a quality mark on the treated wood (Fig. 104). This mark provides a substantial amount of information, including whether or not the treated wood is suitable for ground contact. Where appearance is critical, the quality mark may be omitted. In such cases, the dealer should provide a certificate giving the information indicated in Figure 104. Quality marks or certificates give assurance of proper treatment.

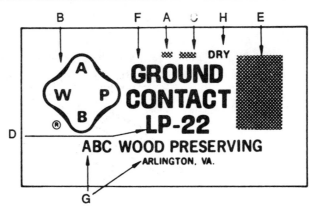

A Year of treatment
B American Wood Preservers Bureau trademark
C The preservative used for treatment
D The applicable American Wood Preservers Bureau quality standard
E Trademark of the agency supervising the treating plant
F Proper exposure conditions
G Treating company and plant location
H Dry or KDAT if applicable

Figure 104—Sample AWPB quality mark.

Treated wood is toxic. Therefore, be sure to ask your supplier for the Safety Data Sheet approved by the EPA, which will tell how to use and handle the material properly. In particular, wood treated with creosote or pentachlorophenol should not be used within the house.

Another way to add just a little extra protection to wood if it becomes wet periodically is on-site brush or dip treatments. In these cases, the wood should be thoroughly dried and then liberally brushed with or totally submerged in the preservative. Most retail lumberyards and paint stores have preservative solutions available to the general public. Remember, the more preservative the wood soaks up and the deeper it penetrates, the better the protection will be. However, brush treatments are only superficial and will not protect wood from decay under severe conditions. Dip treatments do not provide much additional protection except for short pieces.

Some pest control operators can solve moisture and wood decay-related problems. If extensive work is necessary, be sure to check the company's references first.

Insect Damage

If wood has been damaged by insects, first determine if the insects are still active. If so, they should be controlled. Because it is often necessary to use insecticides or fumigants to control wood-infesting insects, consult a licensed pest control operator for help. If the damage is excessive, some wood members may need to be replaced. If there is a possibility of reinfestation, pressure-treated wood should be used.

Rodent Damage

Numerous vertebrate animals, such as rodents, bats, snakes, and raccoons may also infest houses. If a solution is not obvious, consult a pest control operator. Also, the county cooperative extension agent will usually have literature or suggestions on how to correct the problem.

Siding

Many choices of wood-based materials, masonry veneers, and metal or vinyl sidings are available for covering exterior walls. Wood siding comes in several patterns and can be finished naturally, stained, or painted. Wood shingles, plywood, and hardboard are other types of wood and wood-based exterior siding. Coatings and films (overlays) applied to base materials, or certain sidings themselves, eliminate the need for refinishing for many years.

Many older homes were sided with wood. The most common problem with these homes is paint failure that is attributed to moisture moving through the wall because of excess water washing down the wall or the absence of a vapor retarder; the wood siding itself is not the cause of the problem. Correction of these problems is discussed in "Exterior Finishes" in the "Rehabilitation" section.

On some homes, a few pieces of siding may need to be replaced. On others, it may be desirable to re-side the entire structure. If the entire house is being re-sided, consider installing rigid board or other types of insulation under the new siding materials. Insulation is discussed in "Thermal Protection" and "Moisture Control" in the "Rehabilitation" section. If insulation is used under the siding, the nails must be longer than the sizes given in the following sections to ensure that they will hold adequately.

Replacement
Techniques

Any conventional siding material can be used for rehabilitation, but some may be better suited than others. Panel-type siding is probably one of the simplest siding materials to install and one of the most versatile. It can be applied over most surfaces and will help to smooth out unevenness in the existing wall.

If new horizontal wood or nonwood siding is used, the old siding should be removed. Vertical board and panel-type siding may be successfully applied over old siding.

The main difficulty in applying new siding over existing siding is in adjusting the window and door trim to compensate for the added wall thickness. The window sills on most houses extend far enough beyond the original siding so that new siding should not affect them; however, the casing may be nearly flush with the siding and may require some type of extension. One way to extend the casing is to add an additional trim member over the existing casing (Fig. 105). When this addition is made, a wider drip cap may also be required. The drip cap could be replaced, or it could be reused with blocking to hold it out from the wall a distance equal to the new siding thickness (Fig. 106). Another method of extending the casing is adding a trim member to the edge of the existing casing perpendicular to the casing (Fig. 107). A wider drip cap will also be required.

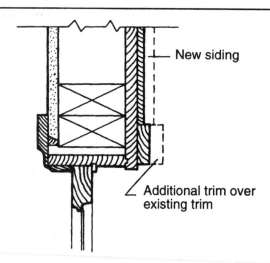

Figure 105—Top view of window casing extended by adding trim over existing trim.

148

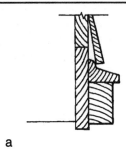

a

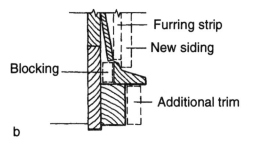

Furring strip

New siding

Blocking

Additional trim

b

Figure 106—Changes in drip cap with new siding: (a) existing drip cap and trim, (b) drip cap blocked out to extend beyond new siding and added trim.

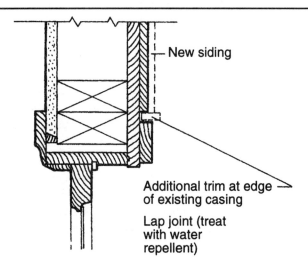

New siding

Additional trim at edge of existing casing

Lap joint (treat with water repellent)

Figure 107—Top view of window casing extended for new siding by adding trim at edge of existing casing.

Exterior door trim can be extended by the same technique used for the window trim.

Any trim added to windows or doors should be liberally treated with a water-repellent preservative as discussed in "Exterior Finishes" in the "Rehabilitation" section. This will help prevent decay and ensure a longer service life if the trim is painted.

Panel Siding

Panel-type siding is available in the form of plywood, hardboard, and particleboard, as well as numerous nonwood materials. The most popular panel sidings are plywood and hardboard. The kind of plywood used depends on the desired finished surface. Some siding materials resist the passage of water vapor, and, when used, should be accompanied by a well-installed vapor retarder applied on the warm side of the insulated walls.

Plywood panel siding is available in a variety of textures and patterns. Sheets are 4 ft wide and often are available in lengths of 8, 9, and 10 ft. Rough-textured plywood is recommended for exterior siding and is well-suited for a natural or rustic finish. Smooth-surfaced plywood siding can be painted with an acrylic latex paint or can be stained; it will not absorb as much stain as rough-textured plywood, so, the finish will not last as long. Most textures can be purchased with vertical grooves (reverse board and batten). The most popular spacings of grooves are 2, 4, and 8 in. Battens are often used with plain panels. They are nailed over each joint between panels and can be nailed over each stud to produce a board-and-batten effect.

Paper-overlaid plywood (called medium-density overlay or MDO) is a particularly good substrate for a paint finish. The paper overlay not only provides a very smooth surface, but also minimizes expansion and contraction from moisture changes.

In new construction, plywood applied vertically and directly over framing may be 3/8 or 1/2 in. thick for 16-in. stud spacing and must be 1/2 in. thick for 24-in. stud spacing. Grooved plywood is normally 5/8 in. thick with about 1/4-in.-deep grooves. Thinner plywood can be installed over existing siding or sheathing; however, most of the available plywood siding will be in these thicknesses. Some sheet or panel materials can be applied horizontally; horizontal joints should be protected by a zee-flashing in accordance with the manufacturer's instructions.

Nail the plywood around the perimeter and at each intermediate stud with galvanized or other rust-resistant nails spaced 7 to 8 in. apart. Nails must be longer than those used for applying siding directly to studs. Effective nail penetration into the wood sheathing and stud should be at least 1-1/2 in. When installing plywood in sheet form, allow a minimum 1/16-in. edge and end-spacing between panels for expansion. If battens are used over the joint and at intermediate studs, nail them with 8d galvanized nails spaced 12 in. apart. Longer nails may be necessary to penetrate thick existing siding or sheathing. Nominal 1- by 2-in. battens are commonly used.

Some plywood siding has shiplap edges. These edges should be treated with a water-repellent preservative, and the siding should be nailed at each side of the joint (Fig. 108a). Square-edge butt joints between plywood panels should be caulked with a sealant (Fig. 108b) with the plywood nailed at each side of the joint.

If the existing siding on gable ends is flush with the siding below the gable, some adjustment will be required to have the new siding at the gable extend over the siding below. This is accomplished by using furring strips on the gable (Fig. 109). Furring must be the same thickness as the new siding applied below.

Nail a furring strip over siding or sheathing to each stud and apply the siding over the furring strips in the same manner as if applying it directly to the studs.

Plywood siding can be special-ordered with factory-applied coatings that are relatively maintenance-free. Although the initial cost of these products is higher than that of uncoated plywood, savings in maintenance may compensate for the extra cost. Such coated siding is usually applied with special nails or other connectors in accordance with the manufacturer's instructions.

Hardboard siding is also available in panels 4 ft wide and up to 16 ft long. It is usually 1/4 in. thick but may be thicker when grooved. Hardboard is usually factory-primed, and finish coats of paint are applied after installation. Hardboard is applied in the same manner as plywood.

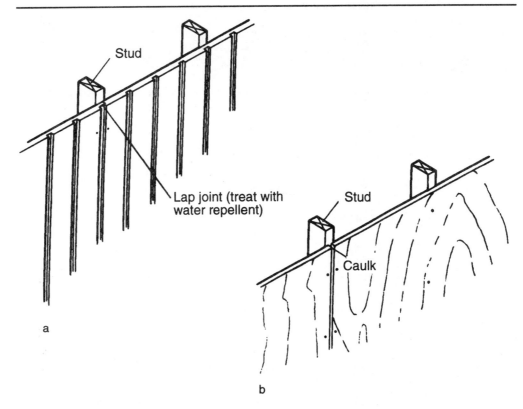

Figure 108—Joint of plywood panel siding: (a) shiplap joints, (b) square-edge joint.

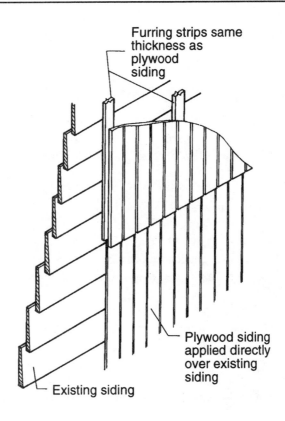

Figure 109—Application of plywood siding at gable end.

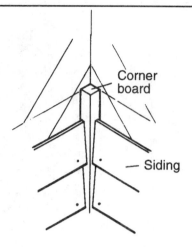

Figure 110—Corner board for applying horizontal siding at interior corner.

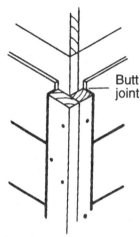

Figure 111—Corner boards for applying horizontal siding at exterior corner.

Corners are finished by butting the panel siding against corner boards, as shown for horizontal sidings (Figs. 110 and 111). Use a 1-1/8- by 1-1/8-in. corner board at interior corners and 1-1/8- by 1-1/2- and 2-1/2-in. boards at outside corners. Apply caulking where siding butts against corner boards, windows, or door casings, and trim boards at gable ends.

Horizontal Wood Siding

Beveled wood siding has been one of the most popular sidings for many years. It is available in 4- to 12-in. widths. The rougher sawn face is exposed if a stain finish is planned. The smooth face can be exposed for either paint or stain. Siding boards should have a minimum of 1-in. lap between adjacent courses. The exposed face should be installed so that the butt edges coincide with the bottom of the window sill and clear the top of the drip cap of window frames (Fig. 79); this requires careful planning.

When siding returns against a roof surface such as at a dormer, it should clear the roof surface by about 2 in. (Fig. 112). Siding cut and installed tightly against the shingles retains moisture after rains and usually results in peeling paint and eventual decay. Shingle flashing that extends well up on the dormer wall will resist the entry of rain. A water-repellent preservative should be used on the siding at the roof line; the ends should be liberally treated.

Bevel siding must be applied over a smooth surface. If the old siding is left on, it should either be covered with panel sheathing or have furring strips nailed over the old siding at each stud. Nail siding at each stud with a galvanized siding nail or other corrosion-resistant nail. Use

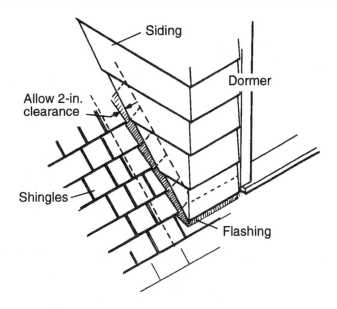

Figure 112—Flashing and siding clearance at dormer walls.

6d nails for siding less than 1/2 in. thick and 8d nails for thicker siding. The nails should clear the top edge of the siding course below (Fig. 113). Butt joints should be avoided where possible, but when required, the joints should be over a stud. Finish interior corners by butting the siding against a corner board, the thickness of which depends on the thickness of the siding (Fig. 110). Exterior corners can be mitered, butted against corner boards 1-1/8 in. thick or more and 1-1/2 and 2-1/2 in. wide (Fig. 111), or covered with metal corners.

Drop siding is another type of horizontal siding that has tongue-and-groove or shiplap edges and can be obtained in 1- by 6-in. and 1- by 8-in. nominal sizes. Actual face widths are 5-1/4 in. and 7-1/4 in., respectively. Drop siding is commonly used for unconditioned buildings and for garages, usually without sheathing. If wide drop siding is used, vertical grain is desirable to reduce shrinkage. With tongue-and-groove siding, the correct moisture content at the time of installation is particularly important because the siding may shrink to a point where the tongue becomes exposed or even totally withdrawn from the groove.

Application methods for drop siding are shown in Figure 113. One or two 8d nails should be used at each stud crossing, depending on the width. Two nails are used for widths greater than 6 in.

Hardboard lap siding is also available, both primed and prefinished, in various widths. Nails with color-matched heads may be available. Install according to the manufacturer's instructions for spacing, nailing, and finishing.

If rigid foam, gypsum, or ordinary (not nail-base) fiberboard sheathing is applied under the siding, the nail lengths must be adjusted to account for the thickness of the sheathing. Guidelines from the American Forest and Paper Association deal with nailing wood-bevel siding and hardboard lap siding over rigid foam sheathing. For 1/2-in. wood-bevel siding installed over 1/2-in. rigid foam sheathing, a 9d (2-3/4-in.) smooth shank or a 7d (2-1/4-in.) ring-shank wood-siding nail is recommended. If 3/4-in. rigid foam sheathing is used, the nail sizes should be increased to a 10d (3-in.) smooth shank or 8d (2-1/2-in.) ring shank. If 3/4-in. wood-bevel siding is installed over 1/2-in. rigid foam sheathing, the

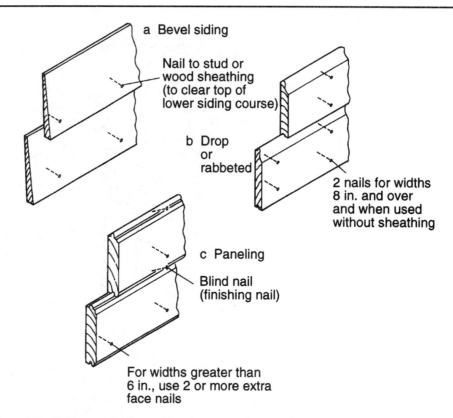

a Bevel siding

Nail to stud or
wood sheathing
(to clear top of
lower siding course)

b Drop
or
rabbeted

2 nails for widths
8 in. and over
and when used
without sheathing

c Paneling

Blind nail
(finishing nail)

For widths greater than
6 in., use 2 or more extra
face nails

Figure 113—Nailing wood siding: (a) bevel-pattern siding, (b) drop-pattern siding, (c) paneling-pattern siding.

wood-siding nail sizes recommended are 10d smooth shank or 8d ring shank. If 3/4-in. rigid foam sheathing is used, the nail sizes should be increased to 12d (3-1/4-in.) smooth shank or 9d ring shank. For 7/16-in. hardboard lap siding installed over either 1/2- or 3/4-in. rigid foam sheathing, a 10d smooth shank hardboard siding nail is recommended.

Vertical Wood Siding

Vertical wood siding is available in a variety of patterns. Probably the most popular is matched (tongue-and-groove) boards. Vertical siding can be nailed directly to 1-in. sheathing boards or to 5/8- and 3/4-in. plywood. Furring strips must be used over thinner plywood because the plywood itself does not have sufficient nail-holding capacity. If the existing sheathing is thinner than 5/8 in., apply 1- by 4-in. nailers spaced 16 to 24 in. apart horizontally, and then nail the vertical siding to the nailers. Blind-nail through the tongue at each nailer with galvanized 7d finish nails. If boards are nominal 6 in. or wider, also face-nail at midwidth with an 8d galvanized nail (Fig. 114). Vertical siding can be applied over existing siding by nailing through the siding into the sheathing.

Another popular vertical siding consists of various combinations of boards and battens (Fig. 115). This type of siding must also be nailed to a thick sheathing or to 2- by 4-in. horizontal nailers installed 16 to 24 in. on center between the studs. The first board or batten should be nailed with one galvanized 8d nail at center. For wide boards, use two nails spaced 1 in. on each side of the center; close spacing is important to prevent splitting if the boards shrink. The top board or batten is then nailed with 12d nails; be careful to miss the underboard and nail only through the space between adjacent boards (Fig. 115). Use only corrosion-resistant nails. Galvanized nails are not recommended for some materials; be sure to follow the siding manufacturer's instructions.

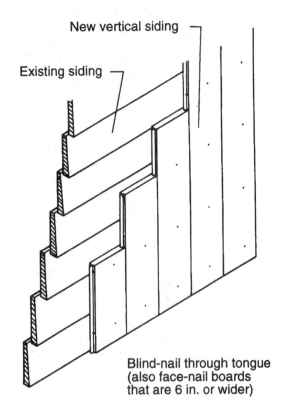

New vertical siding

Existing siding

Blind-nail through tongue
(also face-nail boards
that are 6 in. or wider)

Figure 114—Applying vertical siding.

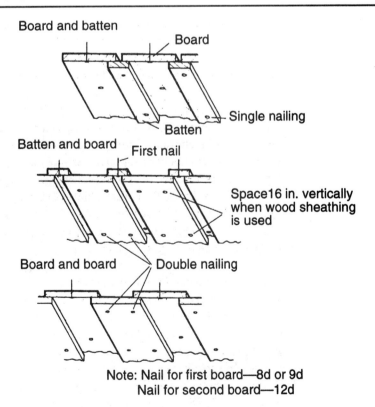

Board and batten

Board

Single nailing

Batten

Batten and board   First nail

Space 16 in. vertically
when wood sheathing
is used

Board and board   Double nailing

Note: Nail for first board—8d or 9d
Nail for second board—12d

Figure 115—Applying board-and-batten siding.

Because wide boards are prone to cupping with changes in moisture content, they should be installed with the bark side turned in (check the orientation of annual rings). This method assures that the concave side is turned inward should there be cupping.

Some older houses have a unique pattern of siding, such as tongue-and-groove or bevel with a bead run along one edge. Many local millwork shops are capable of reproducing these patterns. Requests for small amounts can be expensive because knives used to reproduce the pattern are custom-made.

Wood Shingle and Shake Siding

Some architectural styles may be well-suited to the use of shingles or shakes for siding. These materials give a rustic appearance and can be finished or left unfinished, if desired, to weather naturally. They may be applied in single or double courses over wood or plywood sheathing. If shingles are applied over uneven siding or over a nonwood sheathing, use 1- by 3- or 1- by 4-in. wood nailing strips applied horizontally as a base. The spacing of the nailing strips will depend on the length and exposure of the shingles. Apply the shingles with about 1/8- to 1/4-in. spaces between adjacent shingles to permit expansion during rainy weather. The single-course method consists of laying one course over the other in a manner similar to installing lapsiding. Second-grade shingles can be used because only one-half or less of the butt portion is exposed (Fig. 116).

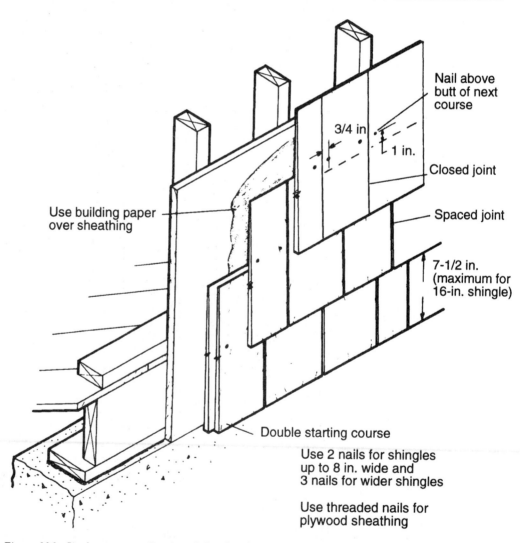

Figure 116—Single-course application of shingle siding.

The double-course method of laying shingles consists of applying an undercourse and nailing a top course directly over it with a 1/4- to 1/2-in. projection of the butt over the undercourse shingle (Fig. 117). With this system, less lap is used between courses.

The undercourse shingles can be lower quality, such as third grade or undercourse grade. The top course should be first grade because of the shingle length exposed. Regardless of the method of applying the shingles, all joints must be offset so that the vertical joints between the top course shingles are at least 1-1/2 in. from the nearest undershingle joints.

Shakes are applied like shingles and are usually available in several types, the most popular being the split-and-resawn. The sawed face is used as the back face. Butt thickness varies from 3/4 to 1-1/2 in.

Recommended exposure distances for shingles and shakes are given in Table 11.

Use galvanized or other corrosion-resistant nails for all shingle applications. Shingles up to 8 in. wide should be nailed with two nails. Secure wider shingles with three nails. Threepenny or 4d galvanized "shingle" nails are commonly used for single-coursing. Galvanized nails with small, flat heads are commonly used for double-coursing where nails are exposed. Use 5d for the top course and 3d or 4d for the undercourse. If plywood sheathing less than 3/4 in. thick is

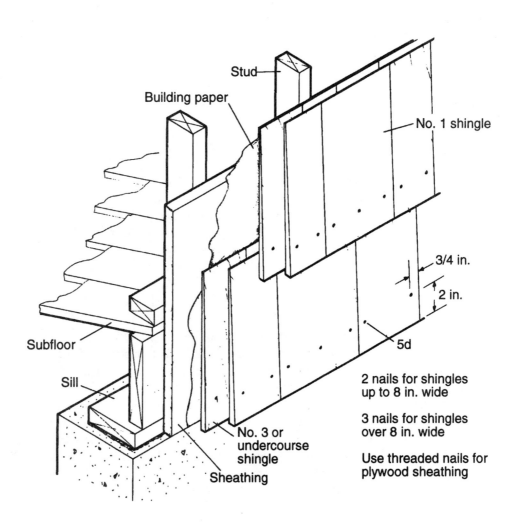

Figure 117—Double-course application of shingle siding.

Table 11—Exposure distances for wood shingles and shakes on side walls

Material	Length of material (in.)	Single coursing	Double coursing	
			No. 1 grade	No. 2 grade
Shingles	16	7-1/2	12	10
	18	8-1/2	14	11
	24	11-1/2	16	14
Shakes	18	8-1/2	14	–
(handsplit	24	11-1/2	20	–
and resawn)	32	15	–	–

used, threaded nails are required for sufficient holding power. Nails should be located 3/4 in. from the edge, and they should be 1 in. above the horizontal butt line of the next higher course in single-course applications, and 2 in. above the bottom of the shingle or shake in double-course applications.

Aluminum and Vinyl Siding

Aluminum and vinyl siding can be purchased in many different patterns and qualities, and requires little maintenance beyond periodic cleaning. Install the siding according to the manufacturer's instructions.

Stucco Finish

Stucco finishes are applied over a coated expanded metal lath and usually over some type of sheathing. If local building regulations permit, a stucco finish can be applied to metal lath fastened directly to the braced wall framework. Waterproof paper should be installed over the studs before the metal lath is applied.

Masonry Veneer

If brick or stone veneer is used as siding, mortar may become loose and crumble, or uneven settlement may cause cracks. In either case, new mortar should be applied both to keep out moisture and to improve appearance. Repair is much the same as that for masonry foundations, except that more attention to appearance is required. After removing all loose mortar and brushing the joint to remove dust and loose particles, dampen the surface. Then apply mortar and tamp it well into the joint for a good bond. Joints should be pointed to conform to existing joints. Exercise particular care to keep mortar off the face of the brick or stone unless the veneer is to be painted. Figure 118 shows a typical brick veneer wall.

Many older houses were built with soft bricks and porous stone trim. After repairing the cracks and mortar joints in the walls of these homes, the entire surface may require treatment with transparent waterproofing. Painted, stained, or dirty brick and stone can be restored to its original appearance by sandblasting. It can then be repainted or waterproofed. Be cautious when sandblasting soft stone and brick because too much material may be removed.

Material Transition

Different materials involving different application methods may be used in the gable ends and in the walls below. Good drainage is necessary at the juncture of the two materials. For example, if vertical boards and battens are used at the gable end and horizontal siding below, a drip cap or similar molding could be installed at the transition (Fig. 119). Flashing should be used over and above the drip cap so moisture will clear the gable material. As an alternative, good drainage can be provided by extending the plate and studs of the gable end out from the wall a short distance or by using furring strips to project the gable siding beyond the wall siding (Fig. 120).

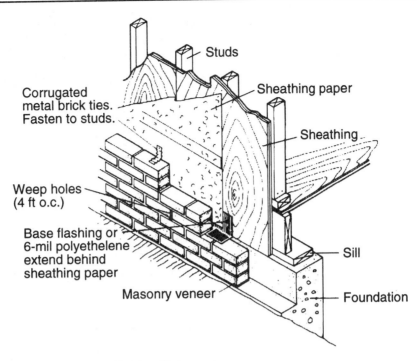

Figure 118—Masonry veneer siding installation.

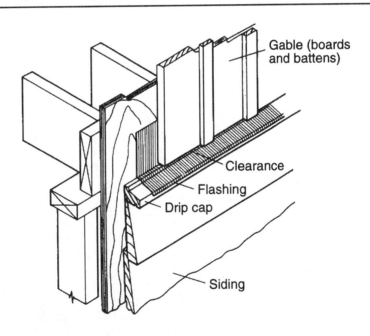

Figure 119—Siding transition at gable end.

Exterior Finishes

Exterior finishes protect the wood surface from weathering and enhance the appearance and architectural style of the house. Proper selection and application of finishes is imperative for a long service life. Some wood properties that affect finish durability are discussed in "Siding" in the "Condition Assessment" section. If finishing problems develop, they must be corrected before a new finish is applied or it will fail also.

Types of Exterior Wood Finishes

Finishes are applied to exterior wood surfaces for various reasons. The particular reason will determine the type of finish selected and subsequently the amount of protection provided to the wood surface as well as the life expectancy for the finish. Finishes can be divided into two

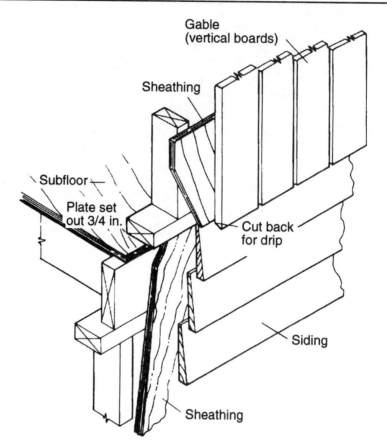

Figure 120—Gable-end siding projection to form drip edge without flashing.

general categories. The first are the opaque (or pigmented) coatings, such as paints and solid-color stains. Virtually all older houses were either painted or left to weather naturally. The second category of finishes includes water repellents, water-repellent preservatives, and semitransparent penetrating stains, which will provide a natural finish. Newer houses may be either painted or given a natural finish.

Paints

Paints are commonly used on wood and provide the most protection against water and against surface erosion caused by weathering. They are also used for aesthetic purposes and to conceal certain defects. Paints contain substantial quantities of pigments, which account for the wide range of colors available. Paints perform best on smooth, edge-grained lumber of lightweight species such as redwood and cedar. Paints are applied to the wood surface and do not penetrate it deeply. Rather, a surface film forms, completely obscuring the wood grain. This surface film can blister or peel if the wood is wetted or if inside water vapor moves through the house wall and into the wood siding because there is no vapor retarder. Original and maintenance costs are often higher for a paint finish than for a water-repellent preservative or penetrating stain finish.

Oil-based paint films provide the best shield from liquid and water vapor. Therefore, they retard penetration of outside moisture and reduce discoloration by wood extractives, peeling of paint from outside moisture sources, and checking and warping of the wood. However, they are not necessarily the most durable. No matter how well sealed, wood still moves with changes in seasonal humidity, thus stressing and eventually cracking the paint because oil-based paints become brittle over time. Latex paints, particularly the acrylics, remain more flexible with age. Even though they allow more water vapor to pass through, they hold up better by stretching and shrinking with the wood. It is important to realize that paint is not a preservative and will not prevent decay if the wood reaches and remains at a high moisture content.

Most complaints about paint are the result of using low-quality products, indicating that better quality paints are usually worth the extra money. Better quality paints normally contain 50 percent solids by weight; paints with less than that may be cheaper by the gallon but more expensive per pound of solids, and more or heavier coats must be applied to achieve equal coverage. *Consumer Reports* (256 Washington St., Mount Vernon, NY 10550) publishes articles from time to time on extensive weather-testing by brand name.

Solid-Color Stains

Solid-color stains (also called hiding, heavy-bodied, opaque, or blocking stains) are opaque finishes that come in a wide range of colors and are essentially thin paints. Solid-color stains are made with a much higher concentration of pigment than the semitransparent penetrating stains (discussed later), but somewhat less than that of standard paints. As a result, they will obscure the natural wood color and grain, and can be applied over old paints or stains. However, the surface texture of the substrate is retained and a flat-finish appearance normally results. Both oil-based and latex solid-color stains form a thin film much like paint and, as a result, can peel from the substrate. The acrylic-based versions are generally the best of all the solid-color stains; however, they may not prevent the bleeding of wood extractives. Solid-color stains are often used on textured surfaces and panel products such as hardboard and plywood.

Natural Wood Finishes

When rehabilitating some older structures, it may be desirable to develop a natural look on the exterior. To some, a natural look means rough, gray, and weathered. This is nature's natural finish. To others, a truly successful natural exterior wood finish is one that will retain the original, attractive appearance of wood with the least change in color and the least masking of wood grain and surface texture. In this case, the finish should inhibit the growth of mildew, protect against moisture and sunlight, and not change the surface appearance or color of the wood. Unfortunately, if wood is left completely unprotected, the grain raises and the wood checks, the checks develop into cracks, boards cup and warp, and mildew and decay can develop. Therefore, the application of a natural finish is usually desirable.

Natural finishes can be divided into two categories. The first category includes the penetrating types such as transparent water repellents, water-repellent preservatives, and semitransparent and pigmented oil-based stains; the second includes film-forming types such as varnishes.

Oils such as linseed, tung, and creosote have been used as natural wood finishes for many years. Linseed oil and tung oil are food sources for mildew, and any finish containing these oils should contain a mildewcide. These oil finishes will normally last 1 to 2 years before they need refinishing.

Water repellents are simply water-repellent preservatives with the preservative left out. They are not good natural finishes but are often applied to wood before painting. They help to stabilize the wood and prevent paint from peeling from the ends and edges of the pieces.

A water-repellent preservative may be used as a natural finish. It reduces warping and checking, prevents water staining at the edges and ends of wood siding, and controls fungal growth. Water-repellent preservatives contain a fungicide, a small amount of wax as a water repellent, a resin or drying oil, and a solvent such as turpentine or mineral spirits. Some waterborne formulations are also available. The wax reduces the absorption of liquid water by the wood (Fig. 121), and the fungicide prevents wood from darkening (graying) by inhibiting the growth of mildew, stain, and decay organisms. Although water-repellent preservatives do not contain coloring pigments, the resulting finish will vary in color depending upon the wood color itself, but will usually weather to a clean, golden tan.

Semitransparent oil-based penetrating stains are moderately pigmented water repellents or water-repellent preservatives. These stains penetrate the wood surface to a certain extent, are porous, and do not form a surface film like paint. Thus, they do not totally hide the wood grain and will not trap moisture that may encourage decay. As a result, they will not blister or peel

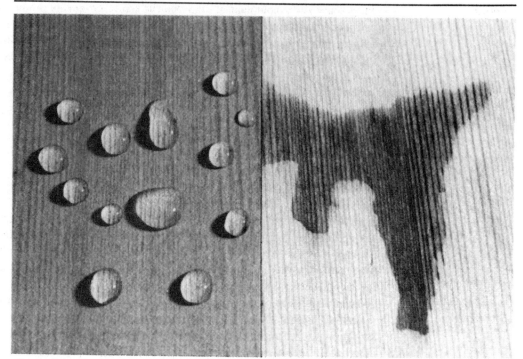

Figure 121—Wood surface brush-treated with water repellent (left) resists penetration by liquid water, whereas the untreated wood surface (right) absorbs water quickly.

even if moisture gets into the wood. Design and construction practices are not as critical for long life as with the film-forming finishes. Penetrating stains are oil-based (or alkyd-based), and some may contain a fungicide or water repellent. The fungicide is important in preventing mildew, especially in southern and coastal areas where moisture is abundant. Latex-based (waterborne) stains are also available, but they do not penetrate the wood surface as do their oil-based counterparts. Thus, they leave a film that may blister or peel.

Semitransparent penetrating stains are most effective on rough lumber and roughsawn plywood surfaces. They also provide moderate performance on smooth wood surfaces but not on smooth plywood surfaces. These stains are also an excellent finish for weathered wood and flat-grained surfaces of dense species that do not hold paint well. They are available in a variety of colors and are especially popular in the brown or the red earth tones because they give a natural or rustic wood appearance. They are not effective when applied over a solid-color stain or on old paint, and they are not recommended for hardboard and waferboard panel surfaces.

Clear coatings of conventional spar, urethane, or marine varnish, which are film-forming finishes, are not generally recommended for exterior use on wood. Ultraviolet light from the sun penetrates the transparent film and degrades the wood under it. Regardless of the number of coats applied, the finish will eventually become brittle from exposure to sunlight, develop severe cracks, and peel, often in less than 2 years. If a long service life is not required, areas that are protected from direct sunlight by an overhang or porch and areas on the north side of a structure can be finished with exterior-grade varnish.

Application

The correct application of a finish to a wood surface is as important for durability and good performance as selecting the most appropriate finishing material. All finishes are either brushed, rolled, sprayed, or applied by dipping. The application technique used, the quantity and quality of finish applied, the surface condition of the substrate, and the weather conditions existing at the time can substantially affect the life expectancy of the finish. The different methods of applying finish, along with other important variables, are discussed later. In addition, manufacturers' directions and safety precautions should always be followed.

162

Proper care and preparation of the wood surface before applying paint is absolutely essential for good performance. Wood and wood-based products should be fully protected from the weather and from wetting both on the job site and after they are installed. Dirt, oil, and other foreign substances that contaminate the surface must be eliminated. Preliminary research has shown that even a 4-week exposure to the weather of a freshly cut wood surface can adversely affect the adhesion of paint to the wood. It is best to paint wood surfaces as soon as possible, weather permitting, before or after installation. Wood that has weathered badly before painting will have a degraded surface that is not good for painting; therefore paint is more likely to peel from the more degraded areas.

To achieve maximum paint life on a new wood surface, follow these steps:

1. Wood siding and trim should be treated with a paintable water-repellent preservative or water repellent, and all joints and cracks should be caulked. Water repellents protect the wood against the entrance of rain and dew, and thus minimize swelling and shrinking. They can be applied by brushing or dipping. Lap and butt joints and the edges of panel products such as plywood, hardboard, and particleboard should be treated especially well because paint normally fails in these areas first (Fig. 122). Allow at least two warm, sunny days for the water repellent or water-repellent preservative to adequately dry before painting the treated surface. If not enough time is allowed for most of the solvent to dry, the paint applied over it may be slow to dry, it may discolor, or it may dry with a rough surface that looks like alligator hide. If the wood has been dip-treated, allow at least 1 week of favorable drying weather. The small amount of wax in the water-repellent preservative will not prevent proper adhesion of the paint.

2. After either the water-repellent preservative or the water repellent has dried, the wood must be primed. For woods such as redwood and cedar, which have large amounts of dark water-soluble extractives, the best primers are good-quality oil-based, alkyd-based, or stain-blocking acrylic latex-based primer paints. The primer seals in or ties up the extractives so that they will not bleed through the top coat. The primer should also be nonporous to inhibit the penetration

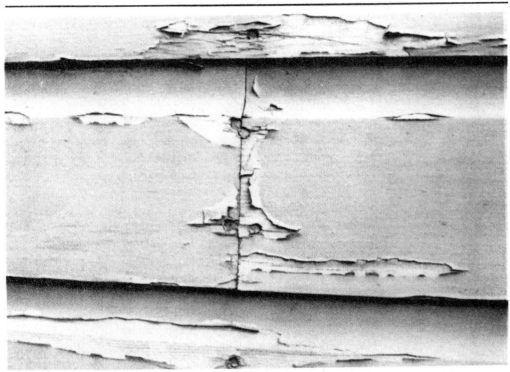

Figure 122—Paint normally fails first around the ends and edges of a board. Liberal application of a water-repellent preservative can prolong the life of paint in these areas.

of rain or dew into the wood surface, thus reducing the tendency of the wood to shrink and swell. A primer should be used whether the top coat is an oil-based or latex-based paint. For species that are predominantly sapwood and free of extractives, such as pine, a high-quality acrylic latex paint may be used as both a primer and top coat. Follow the application rates recommended by the manufacturer. The primer should obscure the wood grain. The top coat should be applied as soon as the primer is dry, about 48 hours for oil-based paints or as recommended by the manufacturer.

3. Two coats of a good-quality all-acrylic latex house paint should be applied over the primer. If it is not practical to apply two top coats to the entire house, two top coats for fully exposed areas on the south and west sides must be considered as a minimum for good protection because these areas are the first to deteriorate. Allow each coat of oil-based paint to cure for 1 to 2 days before applying a second coat; in cold or damp weather, an extra day or two should be allowed. Coats of latex paint can usually be applied within a few hours of each other. On those wood surfaces best suited for painting, one coat of a good house paint over a properly applied primer (a conventional two-coat paint system) should last 4 to 5 years, but two top coats can last up to 10 years.

4. One gallon of paint will cover about 400 ft² of smooth surface area. However, coverage can vary with different paints, surface characteristics, and application procedures. Research has indicated that the optimum thickness for the total dry paint coat (primer and two top coats) is about the thickness of a sheet of newspaper (4 to 5 mil). The coverage of a coat of paint can be checked by applying a pint of paint evenly over a measured area that corresponds to that recommended by the manufacturer. Brush application is always superior to roller or spray application, especially for the first coat. If applying paint by roller or spray, follow by back brushing for optimum performance.

To avoid future separation between paint coats, the first top coat should be applied within 2 weeks of the primer and the second coat within 2 weeks of the first. As certain primer paints weather, they can form a soap-like substance on their surface that may prevent proper adhesion of new paint coats. If more than 2 weeks elapse before applying another paint coat, scrub the old surface with water using a bristle brush or a sponge. If necessary, use a mild detergent to remove all dirt and deteriorated paint. Then rinse well with water and allow the surface to dry before painting. Primer paint left to weather more than 4 weeks should be washed and the surface reprimed.

Oil-based paint may be applied when the temperature is 40°F or above; a minimum of 50°F is desirable for applying latex-based paints. For latex paint films to cure properly, the temperature should not drop below 50°F for at least 24 hours after the paint is applied. Low temperatures cause poor coalescence of the paint film and early paint failure.

Wrinkling, fading, or loss of gloss in oil-based paints and streaking of latex paints can be avoided by not painting in the evenings or cool spring and fall days when heavy dews can form before the surface of the paint has dried thoroughly. Allow at least 2 hours for the paint to dry before sunset. Serious water absorption problems and major finish failure can also occur with some latex paints when applied under these conditions. Likewise, painting in the morning should not begin until after the dew has evaporated.

Solid-Color Stains

Solid-color stains may be applied to a smooth surface by brushing or rolling, but brushing is best. These stains act much like paint. However, solid-color stains are not generally recommended for flat wood surfaces such as decks and window sills. One coat of solid-color stain is adequate, but two coats will provide better protection and longer service. The all-acrylic latex solid-color stains are generally superior to others, especially when two coats are applied.

Unlike paint, lap marks may form with a solid-color stain. Latex-based stains are fast-drying and are more likely to show lap marks than do oil-based stains. To prevent lap marks, follow the procedures suggested under application of semitransparent penetrating stains.

Water-Repellent Preservatives and Water Repellents

The most effective method of applying a water-repellent preservative or water repellent is to dip the entire board into the solution. However, brush treatment is also effective. If wood is treated in place, liberal amounts of the solution should be applied to all lap and butt joints, edges and ends of boards, and edges of panels. Other areas especially vulnerable to moisture, such as the bottoms of doors and window frames, should not be overlooked. One gallon will cover about 250 ft^2 of smooth surface or 100 to 150 ft^2 of rough surface. If these products are used as natural finishes, their life expectancy is only 1 to 2 years, depending upon the wood and the exposure. Treatments on rough surfaces are generally longer lived than those on smooth surfaces. Repeated brushing to the point of refusal (when no more liquid will be absorbed) will enhance the finish durability and performance.

Weathering of the wood surface before applying these materials may be beneficial for both water-repellent preservatives and semitransparent penetrating stains in extending the life of the finish. Weathering opens up checks and cracks, thus allowing the wood to absorb and retain much more of the preservative or stain so the finish is generally more durable.

Semitransparent Penetrating Stains

Semitransparent penetrating stains may be brushed, sprayed, or rolled on. Again, brushing will give better penetration and performance. These oil-based stains are generally thin and runny, so application can be messy. Lap marks will form if stains are improperly applied (Fig. 123). Lap marks can be prevented by staining only a small number of boards or one panel at a time. This method prevents the front edge of the stained area from drying out before a logical stopping place is reached. Working in the shade is desirable because the drying rate is slower. One gallon usually will cover about 200 to 400 ft^2 of smooth surface and from 100 to 150 ft^2 of rough or weathered surface. For long life with penetrating oil-based stain on roughsawn or

Figure 123—Lap marks formed by improper application of semitransparent penetrating stain.

165

weathered lumber, use two coats and apply the second coat before the first is dry. Start on one panel by applying the first coat; move on to another area so that the finish on the first panel can soak into the wood for 20 to 60 minutes. Return to the first panel and apply the second coat before the first has dried. If the first coat dries completely, the second coat cannot penetrate the wood. Finally, about an hour after applying the second coat, use a cloth, sponge, or dry brush that has been lightly wetted with stain to wipe off the excess stain that has not penetrated into the wood. Wiping off the stain that did not penetrate the wood will avoid the formation of an unsightly surface film and glossy spots. Avoid intermixing different brands or batches of stain. Stir stain thoroughly and frequently while applying it to prevent settling and color change.

For the oil-based stains, a two-coat system on rough wood may last as long as 8 years in certain exposures. By comparison, if only one coat is applied on new, smooth wood, its expected life is 2 to 3 years, but succeeding coats will usually last longer.

Caution: Sponges or cloths that are wet with oil-based stain are particularly susceptible to spontaneous combustion. To prevent fires, bury them, immerse them in water, or seal them in an airtight container immediately after use.

Latex semitransparent stains do not penetrate the wood surface but are easy to apply and are less likely to form lap marks. For a long life, use two coats; apply the second coat any time after the first has dried. The second coat will remain free of gloss even on smooth wood. These stains are essentially very thin paints and perform accordingly.

Transparent Coatings

Although short lived, transparent coatings such as polyurethane or spar varnish are used occasionally for exterior applications. The wood surface should be clean, smooth, and dry before application. The wood should first be treated with a paintable water-repellent preservative as discussed under painting procedures. The use of varnish-compatible, durable, pigmented stains and sealers or undercoats will help to extend the life of the finishing system. Finally, at least three top coats should be applied, but the life expectancy for fully exposed surfaces is only 2 years at best. In marine exposures, six coats of varnish should be used for best performance. Varnish built up in many thin coats (as many as six) with a light sanding and with a fresh thin coat added each year will usually perform the best.

Refinishing

Exterior wood surfaces should be refinished only as the old finish deteriorates or for purely aesthetic reasons, such as a change in color or type of finish. However, refinishing too frequently, especially with paint, will lead to a finish buildup and subsequent cracking and/or peeling. In some cases, dirty or discolored paint can be freshened simply by washing it with a mild detergent and water. To achieve maximum service life from a refinished surface, follow the surface preparation and finish application techniques outlined below.

Paint and Solid-Color Stains

Proper surface preparation is essential in refinishing an old paint coat, or a solid-color stain, that has weathered normally. First, scrape away all loose paint. Use sandpaper on any remaining paint to "feather" the edges smooth with the bare wood. Sand the bare wood thoroughly. Then, scrub the bare wood and any remaining paint with a brush or sponge and water. Rinse the scrubbed surface with clean water. Wipe the old paint surface with your hand. If the surface is still dirty or chalky, scrub it again with a detergent. Mildew on bare wood or on the paint should be removed with a diluted solution of household bleach. Rinse the cleaned surface thoroughly with fresh water and allow it to dry thoroughly before repainting. Areas of bare wood should be treated with a water-repellent preservative or water repellent, and allowed to dry for at least two sunny, dry days, then primed. Wipe away any water-repellent preservative accidentally applied on painted areas; paint top coats can then be applied. On redwood and cedar siding, any bare areas should be primed as discussed in "Exterior Finishes." When repainting with oil-based paints, one coat is usually adequate if the old paint surface is still in good condition.

It is particularly important to clean areas protected from sun and rain, such as porches, side walls, and areas under wide roof overhangs. These areas tend to collect dirt and water-soluble materials that interfere with adhesion of the new paint. It is probably adequate to repaint these protected areas every other time the house is painted.

Latex paint can be applied over freshly primed surfaces and on weathered paint surfaces if the old paint is clean and sound. For these situations, first conduct a simple test. After cleaning the surface, repaint a small, inconspicuous area with latex paint, and allow it to dry at least overnight. To test for adhesion, firmly press one end of an adhesive bandage onto the painted surface. Remove the bandage with a jerk or snapping action. If the tape is free of paint, the latex paint is well bonded and the old surface does not need priming or additional cleaning (Fig. 124). If the new latex paint adheres to the tape, the old surface is too chalky and needs more cleaning or an oil-based primer should be applied. The primer should penetrate the old chalky surface and form a firm base for the new paint. If both the latex paint and the old paint adhere to the tape, the old paint is not adaquately bonded to the wood and must be removed before repainting.

Water-Repellent Preservatives and Water Repellents

Surfaces finished with water-repellent preservatives and water repellents can be renewed simply by cleaning them with a bristle brush and applying a new coat of finish. In some cases, mild scrubbing with a detergent followed by rinsing with water is appropriate. The second coat of water-repellent preservative will last longer than the first because more can be applied as it penetrates into the small surface checks that open as the wood weathers.

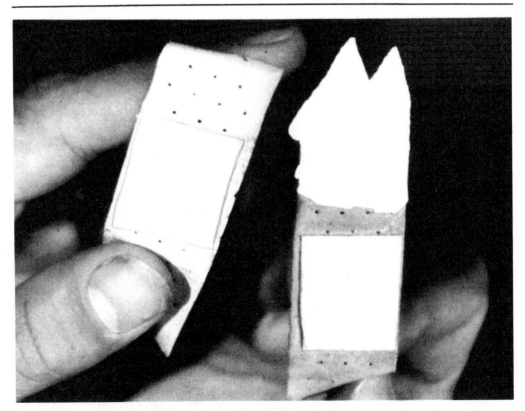

Figure 124—The adhesive bandage test can be used to determine if a new coat of paint is properly bonded to an old surface. The bandage on the left was applied to a well-bonded paint coat. The bandage on the right was pulled off a poorly bonded paint coat.

Semitransparent Penetrating Stains

Semitransparent penetrating stains are also relatively easy to refinish by using a stiff bristle brush to remove all surface dirt, dust, and loose wood fibers. Power-washing can also be used. Then apply a new coat of stain. The second coat of penetrating stain often lasts longer than the first because more can be applied as it penetrates into the small surface checks that open as the wood weathers.

Steel wool and wire brushes should never be used to clean surfaces to be finished with semi-transparent stains or water-repellent preservatives because small iron deposits may be left behind. These deposits can react with certain water-soluble extractives in woods like western redcedar, redwood, Douglas-fir, and the oaks to yield unsightly dark blue-black stains on the surface.

Transparent Coatings

The practices described in "Paints" should be followed for refinishing transparent film-forming finishes such as varnish.

Finish Removal

The removal of paint and other film finishes is a time-consuming and often difficult process. However, it is sometimes necessary if the old surface is covered with severely peeled or blistered paint or if there has been cross-grain cracking from excessive paint buildup. It is also necessary when a penetrating stain or water-repellent finish will be applied to a previously painted surface. There are several ways to remove paint, including sanding, electrically heated blowers or hot plates, blow torches, chemical strippers, sandblasting, and spraying with pressurized water.

Because lead-based paint was used on the exterior of dwellings until 1976, it is best to assume that any exterior paint, including earlier coats that have subsequently been covered and that were applied before this date, contains lead. Because of the potential hazards associated with removing lead-based paint, be sure to review "Lead" in the "Condition Assessment" section and in the "Rehabilitation" section. Re-siding may be considered as an alternative because of the potential health problems and the cost of paint removal.

There are a number of ways to remove nonlead-based paints. Disk or siding sanders are effective in removing old paint, which may be faster than the others discussed here. The depth of cut for the sander can be set with the siding guide, but experienced operators often work freehand. The operator should be careful to remove only paint, not wood. After finishing with the disk sander, smooth the surface somewhat by light hand-sanding or with a straight-line power sander using 120-grit sandpaper along and not across the grain.

Electrically heated paint removers (blowers or hot plate-type) can be used to strip paint. The remover simply heats the paint, causing it to separate from the wood. This usually requires at least a 1,000-watt heater to be effective. Hot-air blower types appear particularly effective.

Liquid paint and varnish removers, such as commercially prepared chemical mixtures, lye, or trisodium phosphate, will also remove paint from wood surfaces. However, after removing the paint with chemical removers, the surface sometimes must be neutralized; before repainting, the wood surface should be sanded in the direction of the grain. Strong caustic solutions, such as lye and trisodium phosphate, may leave the wood surface porous.

Blasting with sand or high-pressure water is another way to remove paint; this usually requires the services of a professional. The sand particles or water can erode the wood as well as strip the paint; softer earlywood is eroded faster than the latewood, resulting in an uneven, rough surface. These rough surfaces may not be suitable for painting.

If a film-type finish must be removed from a structure, consult with local equipment-rental stores and paint dealers for available equipment, or consult professional contractors. To be safe, anyone using sandblasting or water-blasting equipment should wear approved goggles. Dust masks or respirators should be used when sandblasting. Electrical equipment should be double-insulated or equipped with a three-wire grounded outlet. An open-flame blowtorch is another method sometimes used. A blowtorch is effective and inexpensive, but there is a constant danger of starting a fire within the walls of the building from flames that penetrate cracks in the siding. This method is not recommended for wood substrates.

Correction of Finishing Problems

Under normal conditions, paint deteriorates first by soiling or accumulating slight traces of dirt. Next, a flattening stage occurs when the coating gradually starts to chalk and erode away. Paint may also become discolored by mildew, blue stain, wood extractives, pitch, and metals, making repainting necessary. Discoloration problems are eliminated only by identifying and correcting the cause. Refinishing without correcting the cause will result in repeated failure.

Mildew

Mildew is probably the most common finishing problem. It causes a discoloration of house paint and a graying of unfinished wood (Fig. 17). The term mildew applies both to the fungus (a type of microscopic plant life) and to its staining and degenerative effects on paint or wood. The most common species of mildew are black, but some are brown, red, green, or other colors. It grows most extensively in warm, humid climates but is also found in cold northern states. Mildew may be found anywhere on a building whether painted or not, but it is found most commonly on walls behind trees or shrubs where air movement is restricted. Mildew may also be associated with the dew pattern of the house. Dew will form on those parts of the house that are not heated and tend to cool rapidly, such as eaves, the ceilings of carports and porches, and the wall area between studs. The dew provides a source of moisture for mildew growth.

A simple test for the presence of mildew on paint and wood can be made by applying a drop or two of a fresh solution of household liquid bleach (5 percent sodium hypochlorite) to the stained area. The dark color of mildew will usually bleach out in 1 or 2 minutes. Discoloration that does not bleach is probably dirt. It is important to use fresh bleach solution because it deteriorates upon standing and loses its potency.

Some paints are more vulnerable than others to attack by mildew fungi. Zinc oxide, a common paint pigment in top coats, inhibits the growth of mildew, whereas titanium dioxide, another common paint pigment, has very little inhibiting effect. Paints or stains containing linseed oil are very susceptible to mildew; of the available water-based paints, acrylic latex is the most resistant. Porous latex (water-based) paints without a mildewcide, applied over a primer coat with linseed oil, will develop severe mildew in warm, damp climates. With oil-based paints, mildew progresses more readily on exterior flat house paint than on exterior semigloss or gloss enamel.

Mildewcides are poisons for mildew fungi and are often added to paints. The paint label should indicate if a mildewcide is present in the paint. If it is not, it can sometimes be added by the local paint dealer. Paint containing mildewcides, when properly applied to a clean surface, should prevent mildew problems for some time.

To prevent mildew on new wood surfaces in warm, damp climates where it occurs frequently, use a paint containing zinc oxide and mildewcide for top coats over a primer coat that also contains a mildewcide. For mild cases of mildew, use a paint containing a mildewcide. If mildew already exists, it must be killed before repainting or it will grow through the new paint coat. To kill mildew and to clean an area for general appearance or for repainting, a bristle brush or sponge should be used to scrub the painted surface with the following solution:

1/3 cup household detergent
1 quart household bleach (5 percent sodium hypochlorite)
3 quarts warm water

This mixture can also be used to remove mildew from unfinished wood.

Warning: Do not mix liquid household bleach with ammonia or with any detergents or cleaners containing ammonia. Mixed together, the two are a lethal combination that is similar to mustard gas. People have died from breathing the fumes from such a mixture. Many household cleaners contain ammonia, so be extremely careful what type of cleaner is mixed with bleach.

Peeling

Intercoat peeling is the separation of the new paint film from the old paint coat, indicating that the bond between the two is weak (Fig. 125). It usually results from inadequate cleaning of the weathered paint and usually occurs within 1 year of repainting. This type of paint peeling can be prevented by following good cleaning and painting practices.

Intercoat peeling can be caused by allowing too much time between the primer coat and top coat in a new paint job. If more than 2 weeks elapse before a top coat is applied to an oil-based primer, soaplike materials may form on the surface and interfere with the bonding of the next coat of paint. If the period between coats exceeds 2 weeks, the surface should be scrubbed before applying the second coat.

Blistering

Temperature blisters are bubblelike swellings that occur on the surface of the paint film as early as a few hours or as long as 1 to 2 days after painting. Blisters occur only in the last coat of paint (Fig. 126) when a thin, dry skin has formed on the outer surface of the fresh paint and the liquid thinner in the wet paint under the dry skin changes to vapor and cannot escape. When sun shines directly on freshly painted wood, the rapid rise in temperature causes the vapors to expand and produce blisters. Usually only oil-based paints blister in this way. Dark colors that absorb heat and thick paint coats are more likely to blister than white paints or thin coats.

Figure 125—Intercoat peeling of paint is usually caused by poor preparation of the old surface.

To prevent temperature blisters, avoid painting surfaces that will soon be heated. The best procedure is to "follow the sun around the house." The north side of the building should be painted early in the morning, the east side late in the morning, the south side well into the afternoon, and the west side late in the afternoon. However, at least 2 hours should be available for the fresh paint to dry before it cools to the point where condensation could occur. If blistering does occur, allow the paint to dry for a few days, scrape off the blisters, smooth the edges with sandpaper, and spot-paint the area.

Moisture blisters are also bubblelike swellings on the surface of the paint. As the name implies, these blisters usually contain moisture when they are formed. They may occur where outside moisture such as rain enters the wood through joints and other end-grain areas of boards and siding. Moisture also may enter because of improper construction and maintenance practices. Paint blisters caused by outside water are usually concentrated around joints and the end grain of wood, particularly in the lower courses of siding. Paint failure is more severe on the sides of buildings facing the prevailing winds and rain. Damage appears after spring rains and throughout the summer. Moisture blisters may occur in both heated and unheated buildings.

Moisture blisters may also result when water vapor within the house moves to the outside. If the warm side of outside walls does not contain a vapor retarder, the moisture will move through the wall and moisture blisters or paint peeling will result. Such damage is not seasonal and occurs when the faulty condition develops.

Moisture blisters usually involve all coats of paint down to the wood surface. After the blisters appear, they dry out and collapse. Small blisters may disappear completely, but fairly large ones may leave a rough spot, and in severe cases the paint will peel. Thin coatings of new, oil-based paint are the most likely to blister. Old, thick coats are usually too rigid to swell and form blisters, so they will crack and peel instead.

Eliminating the moisture problem and using a vapor retarder are the only practical ways to prevent moisture blisters in paint. The source of the moisture should be identified and eliminated to avoid more serious problems, such as wood decay and loss of insulating value.

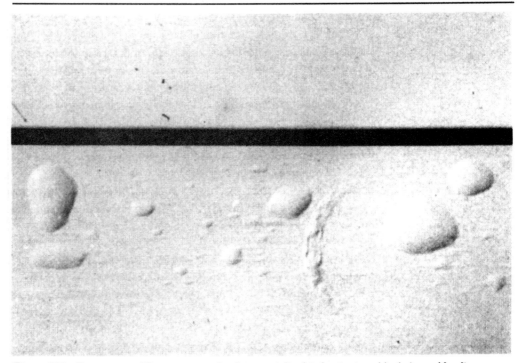

Figure 126—Temperature blisters can result when partially dried paint is suddenly heated by direct sun.

Cross-grain cracking occurs when paint coatings become too thick (Fig. 15), such as on older homes that have been painted several times. Normally, paint cracks in the direction in which it was brushed onto the wood. Cross-grain cracks run across the grain of the wood and paint. Once it has occurred, the only solution is to remove the old paint completely and apply a new finish on the bare wood.

To prevent cross-grain cracking, follow the paint manufacturer's recommendations for spreading rates. Do not repaint unweathered, protected areas such as porch ceilings and roof overhangs as often as the rest of the house. If possible, repaint these areas only as they weather and require new paint. However, if repainting is required, be sure to scrub the areas with a sponge or bristle brush and detergent in water to remove any water-soluble materials that will interfere with adhesion of the new paint. Latex paints, based on either vinyl or acrylic polymers, have not been known to fail by cross-grain cracking.

Discoloration by Water-Soluble Extractives

In some wood species, such as western redcedar and redwood, the heartwood is dark-colored because of the presence of water-soluble extractives. The extractives give these species their attractive color, good stability, and natural decay resistance, but they can also discolor paint. The heartwood of Douglas Fir and Southern Pine can also produce extractive staining problems.

When extractives discolor paint, moisture is usually the culprit. The extractives are dissolved and leached from the wood by water. The water then moves to the paint surface, evaporates, and leaves the extractives behind as a reddish-brown stain. Latex paints and so-called breather or low-luster oil paints are more porous than conventional paints and thus are more susceptible to extractive staining.

Diffused discoloration from wood extractives is caused by the water from rain and dew that penetrates a porous or thin paint coat. It may also be caused by rain and dew that penetrate joints in the siding or by water from faulty roof drainage and gutters. Diffused discoloration from wood extractives is prevented best by following good painting and maintenance practices.

Rundown or streaked discoloration can also occur when water-soluble extractives are present (Fig. 18). This discoloration results when the back of the siding is wetted, the extractives are dissolved, and the colored water runs down from the lap to the face of the adjacent painted siding boards. A rundown discoloration can result from water vapor within the house moving to the exterior walls and condensing during cold weather or by water draining into exterior walls from roof leaks, faulty gutters, and ice dams, or rain and snow blown through louvers in vents.

Rundown discoloration can be prevented by reducing condensation or the accumulation of moisture in the walls. Houses should have a vapor retarder (continuous 6-mil polyethylene sheet, for example) installed on the inside of all exterior walls. If a vapor retarder is not practical, the inside of all exterior walls should be painted with a vapor-resistant paint. Water vapor in the house must be reduced and good construction, maintenance, and attic venting practices must also be followed.

Once the moisture source is eliminated, rundown discoloration will usually weather away in a few months. However, discoloration in protected areas can become darker and more difficult to remove with time. In these cases, discolored areas should be washed with a mild detergent soon after the problem develops. Paint cleaners may also be effective on darker stains.

Chalking

Chalking results when a coat of paint gradually weathers or deteriorates, releasing the individual particles of resin and pigment. These individual particles appear as a fine powder on the paint surface. Most paints chalk to some extent, which is desirable because it allows the paint surface to be self-cleaning. However, chalking is objectionable when it washes down over a surface with a different color (Fig. 14) or when it causes premature disappearance of the paint film through excess erosion. Chalking is also a common cause of fading with colored or tinted paints.

The manner in which a paint is formulated may determine how fast it chalks. Therefore, if chalking is likely to be a problem, select a paint that the manufacturer has indicated will chalk slowly.

When repainting surfaces that have chalked excessively, proper preparation of the old surface is essential if the new coat is expected to last. Scrub the old surface thoroughly with a detergent solution to remove all old chalk deposits and dirt. Rinse thoroughly with clean water before repainting. The use of a top-quality oil-based primer or a stain-blocking acrylic latex primer may be necessary before latex top coats are applied. Otherwise, the new coat may peel. Discoloration or chalk that has run down on a lower surface may be removed by vigorous scrubbing with a good detergent. This discoloration will also gradually weather away if the chalking problem on the painted surface has been corrected.

Iron Stain

Rust is one type of staining associated with iron. When standard ferrous nails are used on exterior siding and then painted, a red-brown discoloration may appear through the paint in the immediate vicinity of the nailhead. To prevent rust stains, use high-quality galvanized, stainless steel, or aluminum nails. The heads of poor-quality galvanized nails can be chipped when they are driven into the siding; these nails corrode easily, and like standard ferrous nails, they cause unsightly staining of the wood and paint. If rust is a serious problem on a painted surface, the nails should be countersunk and caulked and the area spot-primed and then top-coated.

Unsightly rust stains may also occur when standard ferrous nails are used with any of the other finishing systems such as solid-color stains, semitransparent penetrating stains, and water-repellent preservatives. Rust stains can also result when screens and other metal objects that are subject to corrosion and leaching are fastened to the surface of the building (Fig. 127).

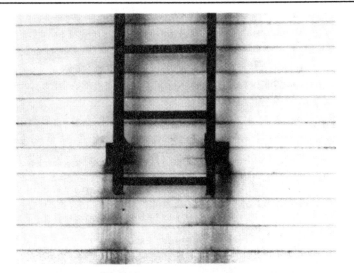

Figure 127—Metal fasteners or window screens can corrode and later discolor paint as leaching occurs.

A chemical reaction with iron can result in an unsightly blue-black discoloration of wood. In this case, the iron reacts with certain wood extractives such as tannins or tannic acid in cedar, redwood, or oak to form the discoloration. Ferrous nails and other iron appendages are the most common source of iron for chemical staining, but problems have also been associated with traces of iron left from cleaning the wood with steel wool, wire brushes, or even iron tools.

A solution of oxalic acid in water will remove the blue-black chemical discoloration if it is not already sealed beneath a finishing system. The stained surface should be given several applications of a solution containing at least 1 lb of oxalic acid per gallon of hot water. After the stains disappear, the surface should be thoroughly washed with warm, fresh water to remove the oxalic acid and any traces of the chemical causing the stain. If all sources of iron are not removed or protected from corrosion, staining may recur.

Warning: Oxalic acid is toxic and should be used with great caution.

Blue Stain

Blue stain is caused by microscopic fungi that commonly infect the sapwood of all woody species and produce a blue-black discoloration of the wood. Wood in service may contain blue stain, but there will not be any detrimental effects as long as the wood moisture content is kept below 20 percent. Wood in properly designed and well-maintained houses usually has a moisture content of 7 to 14 percent. However, if the wood is exposed to moisture such as rain, condensation, or leaky plumbing, the moisture content will increase, the blue-stain fungi will develop further, and decay may follow. To prevent blue stain from discoloring paint, follow good construction and maintenance practices that will keep the wood dry. First, do whatever is possible to keep the wood dry: provide an adequate roof overhang, and properly maintain the shingles, gutters, and downspouts; slope window and door casings out from the house, allowing water to drain away rapidly; use a vapor retarder on the interior side of all exterior walls in northern climates to prevent condensation; vent clothes dryers, showers, and cooking areas to the outside; and avoid overusing humidifiers. Before painting, treat new wood with a water-repellent preservative, then a nonporous mildew-resistant primer, and finally at least one top coat of paint containing a mildewcide should be applied. If the wood has already been painted, remove the old paint and allow the wood to dry thoroughly. Apply a water-repellent preservative, and then repaint as described previously.

An undiluted 5-percent solution of sodium hypochlorite (ordinary liquid household bleach) may sometimes remove blue-stain discoloration, but it is not a permanent cure. Be sure to use a fresh solution of bleach, because its effectiveness can diminish with age. In any event, if the wood has a continuously high moisture content, the source of the moisture must be eliminated.

Brown Stain Over Knots

The knots in many softwood species, particularly pine, contain an abundance of resin. The resin can sometimes cause paint to peel or turn brown (Fig. 128). In most cases, this resin is "set" or hardened by the high temperatures used in kiln-drying of construction lumber.

Good painting practices should eliminate or control brown stain over knots. First, apply a good primer recommended for use over knots to the bare wood. Then follow with two top coats. Do not apply ordinary shellac to the knot area; this may cause early paint failure in outdoor exposure. There are specialty shellacs (pigmented) available as knot sealers.

Exudation of Pitch

Pine and Douglas Fir can exude pitch (resin), while the various cedar species (except western redcedar) can exude oils. Normally pitch and oils are not a problem because lumber manufacturers have learned how to "set" pitch and evaporate excess oil during kiln-drying. The material is simply planed or sanded off later in the manufacturing process and does not present additional problems. However, where the proper schedules are not used in drying the lumber, problems can result (Fig. 129).

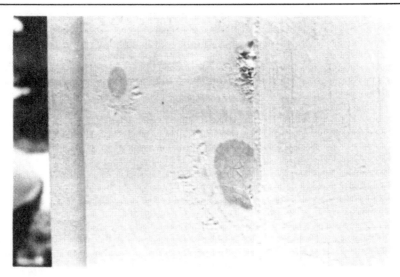

Figure 128—Brown discoloration of paint due to resin exudation from a knot.

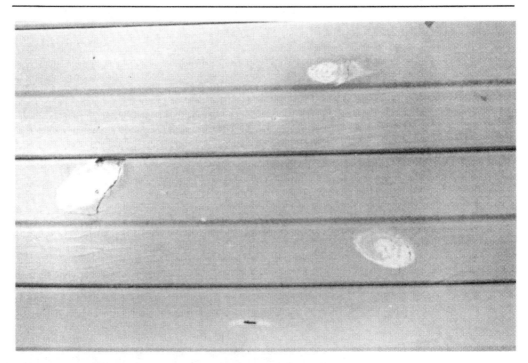

Figure 129—Exuded pitch from knots and pitch pocket on siding.

When exudation occurs before the wood has been painted, it should be removed. If the exuded pitch has hardened, it can be removed fairly easily with a putty knife, paint scraper, or sandpaper; if it is still soft, such procedures smear it over the surface of the wood. If the smeared pitch is allowed to remain on the wood surface, the paint is likely to alligator, crack, and fail early over the affected areas. Exuded pitch that is still soft should be removed by scrubbing it thoroughly with cloths wetted with denatured alcohol and sandpapering the surface after most of the soft pitch has been scrubbed off. Any further exudation that occurs before subsequent coats of paint are applied should be removed by scrubbing with alcohol.

If exudation takes place after painting, the wood might best be left alone until it is time to repaint. The pitch should then be scraped off completely before applying new paint. If a few boards in the structure have proved particularly unsightly because of exudation or because of early paint failure, it may be wise to replace them with new lumber before repainting. In extreme cases, such boards have been known to exude pitch for many years. Exudation is favored by fluctuations in temperature or by warming the wood to high temperatures. Repainting should be deferred until all further exudation has ceased or until repainting has become necessary for other reasons. There are no totally reliable paints and painting procedures that prevent exudation of pitch.

Water Stain

Wood siding, particularly if it is left unfinished or if its natural finish has started to deteriorate, can become water stained (Fig. 130).

Water stain is most common at the base of side walls where rainwater runs off a roof, hits a hard surface, and splashes back onto the side of the building. The water causes the finish to deteriorate faster in this area. If the finish is not replaced, the water can begin to remove the water-soluble extractives, which accelerates the weathering process and results in a water-stained area. Water stain can also be seen where gutters overflow. To prevent water stain, follow good construction practices that keep water from contacting the wood. The wood should be treated regularly with a water-repellent preservative. Removing water stains can be very difficult. Sometimes scrubbing with mild detergent and water is effective. Light sanding can be tried on smooth wood surfaces. Liquid household bleach or oxalic acid solutions have been used with various degrees of success.

Interior Structural Changes

In the layout of a house, it is desirable to have separate areas for various family functions and to provide good traffic circulation through and between them. In a moderate-cost rehabilitation project, however, major modifications may not be practical. Even if cost is not a major limitation, there will have to be compromises in the layout. The considerations presented here are goals to aim for if practical; inability to satisfy them fully should not be an obstacle to restoring the home to a sound, comfortable condition.

Figure 130—Rainwater splash on naturally finished plywood siding. Rainwater runs from the roof and wets siding as it hits the concrete sidewalk.

176

Floor Plan

The layout of the house should be zoned to have three major family functional areas for relaxation, working, and privacy. The relaxation zone includes recreation, entertaining, and dining. In a small house, these may all be performed in one room, but a larger house may have a living room, dining room, family room or den, study, and recreation room. The latter is frequently in the basement. The working zone includes the kitchen, laundry or utility room, and possibly an office or shop. The privacy zone consists of bedrooms and baths; it may also include the den, which can double as a guest bedroom. In some layouts, a master bedroom and bath may be located away from the rest of the bedrooms so the privacy zone is actually in two parts.

Zones within the house should be located for good relationship to outdoor areas. If outdoor living in the backyard is desirable, perhaps the living room should be at the back of the house. The working zone should have easy access to the garage, the dining room, and outdoor work areas. The main entrance to the house should be easily accessible to the driveway or usual guest parking area, which may be on either the front or side of the house. When considering where to locate rooms and entrances, conventional arrangement of the past should not be binding; convenience in the particular situation should govern.

Traffic Circulation

One of the most important items in layout is traffic circulation. Ideally there should be no traffic through any room. This is usually difficult to accomplish in the living and work areas. A more feasible plan is to keep traffic from cutting through the middle of the room. Many older houses have doors centered in the wall of a room; this not only directs traffic through the middle of the room, but also cuts the wall space in half, making furniture arrangement difficult. Study the plan and observe where a door might be moved from the middle of a wall to a corner of the room; however, relocating doors is costly and should be limited in a moderate-cost rehabilitation project. Also consider where doors might be eliminated to prevent traffic through a room. Figure 131 shows an example of layout improved by relocating the doors.

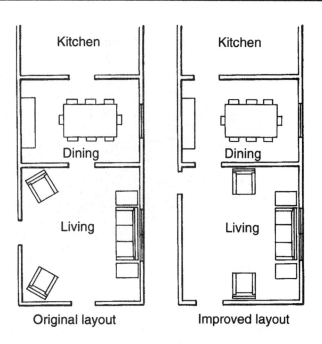

Original layout   Improved layout

Figure 131—Relocation of doors to direct traffic to one side of rooms.

Often rooms are not the desired sizes, and it may be necessary to move some partitions. Such a change is not difficult if the partition is nonload-bearing and plumbing, electrical, or heating services are not concealed within it. It is possible to move load-bearing partitions by adding a beam to support the ceiling from which the partition is removed.

To determine whether or not a partition is load-bearing, check the span direction of ceiling joists. If joists are parallel to the partition, the partition is usually nonload-bearing (Fig. 132); however, it may be supporting a second-floor load. In most constructions where the second-floor joists are perpendicular to the partition, the joists do require support, so the partition is load-bearing (Fig. 132). An exception would be when trusses span the width of the building, making all partitions nonload-bearing.

Although removal of a nonload-bearing partition will not require a structural modification, the wall, ceiling, and floor will require repairs where the partition was removed.

If rooms are small, the owner may wish to remove partitions to give a more open, spacious feeling. A partition between living and dining rooms can be removed to make both seem larger and perhaps make part of the space serve a dual purpose (Fig. 133). In some instances, un-needed bedrooms adjacent to the living area can be used to increase living room or other living space by removing a partition. In other projects, removing a partition between a hallway and a room in the living area provides a more spacious feeling, even though traffic continues through the hallway (Fig. 134).

Window Placement

Window placement influences general arrangement and should also be considered in the floor plan. Moving windows is costly, and it involves changing studs, headers, interior and exterior finish, and trim. The number of windows relocated should be limited in the moderate-cost rehabilitation project, but if the changes are practical, properly placed windows can enhance the livability of a house. If possible, avoid small windows scattered over a wall; they cut up the wall space and make it less usable. Attempt to group windows into one or two areas to leave more wall space undisturbed (Fig. 135). If there is a choice of outside walls for window placement, south walls rank first in cold climates. Winter sun shines into the room through a south window and heats the house, but in the summer, the sun is at a higher angle so that even a small roof overhang shades the window. In extremely warm climates, windows on the north side may be preferable to avoid too much heat gain even in the winter. Windows on the west side should be avoided as much as possible because the late afternoon sun is so low that there is no way of shading the window.

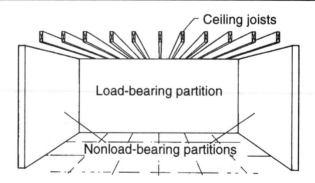

Figure 132—Load-bearing and nonload-bearing partitions. (Second-floor load may place load on any partition.)

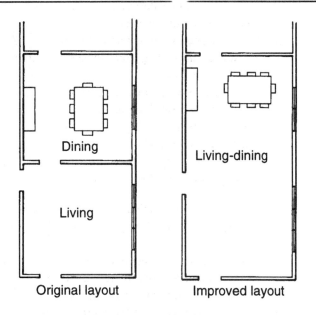

Figure 133—Removal of partition for better space utilization.

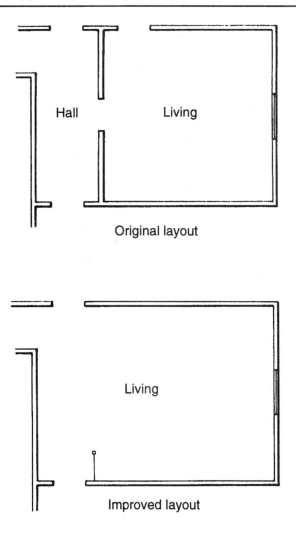

Figure 134—Removal of partition for more spacious feeling.

Spaced windows

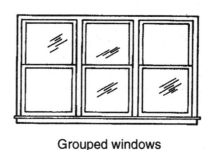

Grouped windows

Figure 135—Grouping improves window placement and available wall space.

Windows have three major functions: they admit daylight, allow ventilation of the house, and provide a view. Points to consider in planning for each of these three functions are discussed here.

To ensure adequate light:

1. Provide glass areas in excess of 10 percent of the floor area of each room.

2. Place principal window areas toward the south, except in warm climates.

3. Group window openings in the wall to eliminate undesirable contrasts in brightness.

4. Screen only those parts of the window that open for ventilation.

5. Mount draperies, curtains, shades, and other window hangings above the head of the window and to the side of the window frame to free the entire glass area when the hangings are open.

To ensure good ventilation:

1. Provide ventilation in excess of 5 percent of the floor area of a room.

2. Locate the ventilation openings to take full advantage of prevailing breezes.

3. Locate windows to effect the best movement of air across the room and within the level where occupants sit or stand. (Ventilation openings should be in the lower part of the wall unless the window swings inward in a manner to direct air downward.)

To provide a good view:

1. Minimize any obstructions in the line of sight when sitting or standing, depending on the use of the room.

2. Determine sill heights on the basis of room use and furniture arrangement.

Closets

An item sometimes overlooked in planning the layout of a house is closets. Most older houses have few if any closets. Plan for a coat closet near both the front and rear entrances. There should be a cleaning closet in the work area and a linen closet in the bedroom area. Each bedroom should have a closet. If a bedroom is large, it is simple to build a closet across one end of the room. It is more difficult to find extra space for closets in a small house. Look for wasted space, such as the end of a hallway or at a wall offset. If the front door opens directly into the living room, a coat closet sometimes can be built beside or facing the door to form an entry (Fig. 136).

In the story-and-a-half house, closets can often be built into the attic space where the headroom is too limited for occupancy.

Ideally, closets used for hanging clothes should be at least 24 in. deep, but shallower closets are also practical. The exception is the walk-in storage closet, which is very useful and should be considered if space is available. If existing closets are narrow and deep, rollout hanging rods can make them more usable. To make the best use of closet space, plan for a full-front opening.

In many situations, plywood wardrobes may be more practical than conventional closets that require studs, drywall, casing, and doors. More usable space can be made by dividing the wardrobe into different spaces for various types of storage and installing the appropriate doors or drawers.

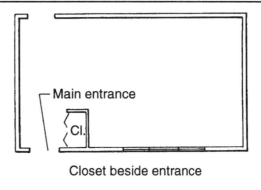

Closet beside entrance

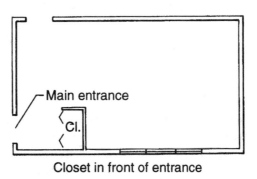

Closet in front of entrance

Figure 136—Entry formed by coat closet.

Porches

Porches on older houses are often very narrow, have sloping floors, and cannot be enlarged easily. They do not lend themselves to the outdoor dining and entertaining that are popular today. The area for this type of activity should be considered when determining whether or not to retain an existing porch and in planning a new porch that will be useful.

Basement
Conversion

An unfinished basement may be one of the easiest places for expansion, although certain conditions must be met if it is to be used for habitable rooms. A habitable room is defined as a space used for living, sleeping, eating, or cooking. Rooms not included and therefore not bound by the requirements of habitable rooms include bathrooms, toilet compartments, closets, halls, storage rooms, laundry and utility rooms, and basement recreation rooms. The average finish-grade elevation at exterior walls of habitable rooms should not be more than 48 in. above the finish floor. The average ceiling height for habitable rooms should not be less than 7 ft 6 in. Local codes should be checked for exact limitations. Other basement rooms should have a minimum ceiling height of 6 ft 9 in.

Dampness in the basement can be partially overcome by installing vapor retarders in the floor and walls if they were not already there. Plan to use a dehumidifier during summer in an extremely damp basement.

One of the main disadvantages of basement rooms is the lack of natural light and view. If the house is on a sloping lot or the lot is graded in a manner to permit large basement windows above grade, the basement is much more usable than in a house where the basement has only a few inches at the top of the wall above grade. However, even a completely sunken basement can have natural light if large areaways are built for windows and the walls of the areaways are sloped so that sunlight can easily reach the windows (Fig. 137). At least one window large enough to serve as a fire exit is recommended and often required by code. It may be convenient to extend the areaway to form a small sunken garden (Fig. 138), but adequate drainage must be included.

The basement's usefulness may be increased by adding a direct outside entrance. This adds to fire safety by giving an alternate exit and it is particularly desirable if the basement is to be used as a workshop or for storing lawn and garden equipment. Try to place an outside stairway under cover of a garage overhang, breezeway, or porch to protect it from ice, snow, and rain. Otherwise, use an areaway-type entrance with a door over it (Fig. 139).

Interior Finishing

Interior finishing differs from exterior finishing mainly in that interior woodwork usually requires much less protection against moisture, but it does require more exacting standards of appearance and cleanability. Good interior finishes should last much longer than paint or other coatings on exterior surfaces. Veneered panels and plywood, however, present special problems because of their tendency to check.

Types of Finishes

Opaque finishes such as paint and clear finishes such as varnish, shellac, or lacquer are the two broad categories of interior finishes. In addition, flat-oil finishes allow the natural grain to show but do not form a surface film of appreciable thickness. This section discusses the types of finishes available and their proper application.

Paint

Interior surfaces may be easily painted by procedures similar to those for exterior surfaces. As a rule, however, smoother surfaces, better color, and a more lasting sheen are demanded for interior woodwork, especially wood trim; therefore, enamels or semigloss enamels are used rather than flat paints. Water repellents or water-repellent preservatives are generally not used, except in areas where water may contact the finish (bathrooms, shower areas, saunas, etc.); check whether these finishes are suitable for interior use.

Figure 137—Large basement window areaway with sloped sides.

Figure 138—Sunken garden forming large basement window areaway.

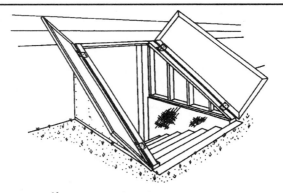

Figure 139—Areaway type of basement entrance.

Before applying an enamel finish, sand the wood surface until it is extremely smooth. Remove the surface dust with a tack cloth.

Enamel finishes accentuate imperfections such as planer marks, hammer marks, and raised grain. For the smoothest surface, it is helpful to sponge softwoods with water, allow them to dry thoroughly, and then sand them lightly with new sandpaper before painting with enamel. Woodwork in new buildings should be allowed adequate time to reach equilibrium moisture content with the atmosphere of the heated building. Depending on the original moisture content of the wood, this may require 1 to 2 months.

Finishing hardwoods with large pores, such as oak and ash, requires that the pores be filled with wood filler (see section on fillers). After filling and sanding, successive applications of interior primer and sealer, undercoat, and enamel are used.

For softwoods, knots in species such as white pine, ponderosa pine, or Southern Pine should be sealed with shellac or a special knot sealer before priming. It is sometimes necessary to apply a coat of pigmented shellac or special knot sealer over the entire surface of white pine and ponderosa pine to retard discoloration of light-colored enamels by colored matter that is present in the resin of the heartwood. One or two coats of enamel undercoat are then applied; this should completely hide the wood and also present a surface that can be sandpapered smooth. For best results, the surface should be sanded just before applying the finish enamel; however, this step is sometimes omitted. After the finishing enamel has been applied, it may be left with its natural gloss or rubbed to a dull finish. If wood trim and paneling are finished with a flat paint, the surface preparation need not be as exacting.

Transparent Finishes

Transparent finishes are used if the natural color and grain of the wood are preferred. Most finishing consists of some combination of sanding, staining, filling, sealing, surface coating, and waxing. Before finishing, planer marks and other blemishes should be removed by sanding.

Stains

Both softwoods and hardwoods are often finished without staining, especially if the wood has a pleasing and characteristic color. Stain often accentuates color differences in the wood surface because the grain pattern absorbs unequally. With hardwoods, such emphasis of the grain is usually desirable.

The most commonly used stains are the "nongrain-raising" ones in solvents that dry quickly. Stains on softwoods color the earlywood more strongly than the latewood, reversing the natural gradation in color unless the wood has been sealed first with a wash coat. Pigment-oil stains, which are essentially thin paints, are less subject to this problem and are therefore more suitable for softwoods. Alternatively, to provide nearly uniform coloring, the softwood may be coated with penetrating clear sealer before any type of stain is applied.

Fillers

Large pores in hardwoods are usually filled after staining and before finishing if a smooth coating is desired. The filler may be transparent and not have an effect on the color of the finish, or it may be colored to contrast with the surrounding wood.

For finishing purposes, hardwoods may be classified as follows:

Hardwoods with large pores	Hardwoods with small pores
Ash	Alder, red
Butternut	Aspen
Chestnut	Basswood
Elm	Beech
Hackberry	Cherry
Hickory	Cottonwood
Lauans	Gum
Mahogany	Magnolia
Mahogany, African	Maple
Oak	Sycamore
Sugarberry	Yellow-poplar
Walnut	

Birch has pores that are large enough to take wood filler effectively when desired but, as a rule, small enough to be finished satisfactorily without filling.

Hardwoods with small pores may be finished with paints, enamels, and varnishes in exactly the same manner as softwoods.

A filler may be a paste or liquid and may be natural or colored. It is applied by brushing first across the grain and then along the grain. Surplus filler must be removed immediately after the glossy wet appearance disappears. To remove excess filler, wipe first across the grain to pack the filler into the pores, then complete the wiping with a few light strokes along the grain. Filler should be allowed to dry thoroughly and sanded lightly before the finish coats are applied.

Sealers

Sealers are thinned varnish, shellac, or lacquer that are used to prevent absorption of surface coatings by the wood; they are also used to prevent some stains and fillers from bleeding into surface coatings, especially lacquer. Shellac and lacquer sealers dry very quickly.

Surface Coats

Transparent surface coatings over the sealer may be varnish, shellac, nitrocellulose or synthetic lacquer, or wax. Coatings of a more resinous nature, especially lacquer and varnish, accentuate the natural luster of some hardwoods and seem to permit the observer to look down into the wood. Shellac and lacquers dry rapidly and form a hard surface, but require more coats than does varnish to build up a luster. Shellac and lacquers are usually not as resistant to water as varnish. Some varnishes and lacquers dry with a highly glossy surface. To reduce the gloss, the surfaces may be rubbed with steel wool. Wax provides protection without forming a thick coating and without greatly enhancing the natural luster of the wood. It is sometimes applied over the finish coat.

Flat-oil finishes, commonly called Danish oils, are also very popular. This type of finish penetrates the wood and does not form a noticeable film on the surface. Two or more coats of oil are usually applied; they may be followed with a paste wax. Such finishes are easily applied and maintained, but they are more subject to soiling than is a film-forming type of finish.

Wood Floors

Wood possesses a variety of properties that make it a highly desirable flooring material for homes. Floor finishes enhance the natural beauty of wood, protect it from excessive wear and abrasion, and make the floors easier to clean. A complete finishing process may consist of four steps: sanding the surface, applying a filler to wood with large pores, applying a stain to achieve a desired color effect, and applying a finish. Detailed procedures and specified materials depend largely on the species of wood used and individual preference for type of finish.

Careful sanding to provide a smooth surface is essential for a good finish because any irregularities or roughness in the wood surface will be magnified by the finish. Development of a top-quality surface requires sanding in several steps with progressively finer sandpaper, usually with a machine unless the area is small. The final sanding is usually done with a 2/0 grade paper. When the sanding is complete, all dust must be removed with a vacuum cleaner and then a tack rag. Steel wool should not be used on floors unprotected by finish because minute steel particles left in the wood may later cause staining discoloration.

A filler is required for wood with large pores, such as oak and walnut, if a smooth, glossy varnish finish is desired.

Stains are sometimes used to obtain a nearly consistent color when individual boards vary too much in their natural color. Stains may also be used to accent the grain pattern. If the natural color of the wood is acceptable, staining may be omitted. The stain should be an oil-based or a nongrain-raising type. Stains penetrate wood only slightly; therefore, the finish should be carefully maintained to prevent wearing through the stained layer. It is difficult to renew the stain at worn spots so that it will match the color of the surrounding area.

Finishes commonly used for wood floors are classified as sealers or varnishes. Sealers, which are usually thinned varnishes, are used widely on residential floors. They penetrate the wood just enough to avoid forming a surface coat of appreciable thickness. Wax is usually applied over the sealer.

Varnish forms a distinct coating over the wood and gives a lustrous finish; if applied amply, it probably will give longer service life.

Floor finishes will last longer if they are kept waxed. Paste waxes generally give the best appearance and durability, and two coats are recommended. More coats of a liquid wax may be necessary to get an adequate film for good performance.

Removal of Existing Finishes

In many older homes, the walls, ceilings, and woodwork may have many coats of finish on them. If the finish is not chipping, peeling, or checking, it can simply be coated again if the surface is adequately prepared first. However, in many cases, the finish has deteriorated to the point where additional finish will not be acceptable. In these cases, the old finish must be removed or the individual members replaced; before doing so, however, review the "Lead" sections under "Condition Assessment" and "Rehabilitation."

If the surface is painted, any of the methods described in "Finish Removal" in the "Rehabilitation" section may be used. However, remember that if a clear or natural finish is to be applied, the wood must not be physically damaged by the removal process.

In many older homes, beautifully grained hardwoods were used for the trim work. Normally, a natural clear finish was applied to this wood first. Paint was applied at a later date. This natural or clear finish generally seals the pores of the wood so the paint coat does not penetrate. With some effort, the paint and old finish can be removed satisfactorily. However, if paint, particularly white, was applied directly over new wood and penetrated the pores, it is nearly impossible to remove it from the pores. Before making a final decision, experiment to determine the kind and condition of the wood below a painted surface.

Surfaces coated with clear finishes such as lacquer, shellac, or varnish often darken with the application of several coats and with age. These surfaces can generally be stripped with commercial products, sanded smooth, and recoated as for new wood. Filling the pores is generally not required because the old finishes probably filled them.

Safety Considerations

Never apply finish or paint remover in a completely closed room or where there is an open flame or fire. Solvent paints and both paint and varnish removers give off fumes that are often flammable and dangerous to breathe. Good cross ventilation not only helps to remove fumes and odors, but it can shorten the drying time.

Some fumes can be especially harmful to infants, children, and canaries and other delicate pets. Avoid sleeping in a freshly finished room until the fumes subside.

When you finish painting or varnishing, dispose of used rags by putting them in a covered metal can. If left lying around, the oily rags could catch fire by spontaneous combustion.

It is recommended that finishes be stored outside of the house. Unless needed for retouching, small quantities of finishes may not be worth saving and should be discarded. If stored, finishes should be kept in a safe, well-ventilated place that is not accessible to children or pets and that is well away from furnaces or other sources of ignition that might cause an explosion. Some finishes cannot withstand freezing.

Vertical Expansion

Additional space is desirable in most homes. When rehabilitating an older home, extra space can often be obtained by finishing the existing attic or modifying the roof to provide more space at the attic level.

Attic Space

A house with a relatively steep roof slope may have attic space that can be used for storage or converted to living area. The attic area can be made accessible for storage by adding a set of fold-down stairs.

If there is already a stairway or if one can be installed, it may be practical to finish the space to form additional bedrooms, a den, a study, a hobby room, or a small, separate living unit. However, if the attic is at third-floor level, local codes may not permit its use for living areas; some codes may require installation of a fire escape.

If a permanent stairway is needed in order to reach the attic area, space problems must be considered. The usual straight-run stairway requires a space 3 ft wide and at least 11 ft long plus a landing at both top and bottom. There must also be a minimum headroom clearance of 6 ft 8 in. along the entire length of the stairs.

If space is limited, consider spiral stairs. Some spiral stairs can be installed in a space as small as 4 ft in diameter; however, check code limitations. Although spiral stairs are quite serviceable, they are somewhat more difficult to ascend or descend and may not be suitable for residents with restricted mobility. Spiral stairs will not accommodate furniture movement, so there must be some other access for moving furniture into the upstairs space.

Attic rooms should have a minimum ceiling height of 7 ft 6 in. over at least one-half the room, and a minimum ceiling height of 5 ft at the outer edges of the room (Fig. 140). The portion of the space with a lower ceiling can be used for bunks, built-in furniture, or storage.

The existing joists that will serve as the attic floor may not be adequate to support a floor load. Consult Table 12 for allowable joist spans or seek professional advice. If the joists are inadequate, a solution may be to double the existing joists.

Insulation, vapor retarders, and ventilation are important when finishing attic space because this space can be particularly hot in the summer. Insulate and install vapor retarders completely around the walls and ceiling of the finished space, and ventilate attic space above and on each side of the finished space. Provide good cross ventilation through the finished space.

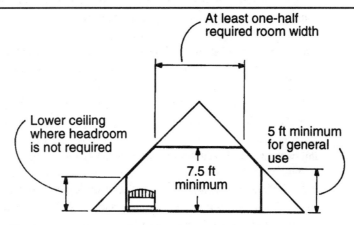

Figure 140—Headroom requirements for attic rooms.

Table 12—Allowable spans for simple floor joists spaced 16 in. on center [a]

	Length of maximum clear span (ft,in.) for lumber with various MOE (x10^6 lb/in^2)[b]										
	1.0	1.1	1.2	1.3	1.4	1.5	1.6	1.7	1.8	1.9	2.0
Living area (40 lb/ft^2 live load)											
Minimum required bending stress (lb/in^2)	920	980	1,040	1,090	1,150	1,200	1,250	1,310	1,360	1,410	1,460
Joist size											
2 by 6	8,4	8,7	8,10	9,1	9,4	9,6	9,9	9,11	10,2	10,4	10,6
2 by 8	11,0	11,4	11,8	12,0	12,3	12,7	12,10	13,1	13,4	13,7	13,10
2 by 10	14,0	14,6	14,11	15,3	15,8	16,0	16,5	16,9	17,0	17,4	17,8
2 by 12	17,0	17,7	18,1	18,7	19,1	19,6	19,11	20,4	21,9	21,1	21,6
Sleeping area (30 lb/ft^2 live load)											
Minimum required bending stress (lb/in^2)	890	950	1,000	1,060	1,110	1,160	1,220	1,270	1,320	1,410	1,360
Joist size											
2 by 6	9,2	9,6	9,9	10,0	10,3	10,6	10,9	10,11	11,2	11,4	11,7
2 by 8	12,1	12,6	12,10	13,2	13,6	13,10	14,2	14,5	14,8	15,0	15,3
2 by 10	15,5	15,11	16,5	16,10	17,3	17,8	18,0	18,5	18,9	19,1	19,5
2 by 12	18,9	19,4	19,11	20,6	21,0	21,6	21,11	22,5	22,10	23,3	23,7

[a]Source: American Forest and Paper Association (*Span Tables for Joists and Rafters*, 1977); other information should be used for other joist spacing.

[b]MOE is modulus of elasticity.

It may be possible to create living space simply by installing finish ceiling, walls, and floor covering. However, in some instances, headroom may be inadequate; often, for example, adequate height exists only along a narrow strip running down the middle of the attic. In other instances, there may be insufficient natural light and/or ventilation. These problems can be solved by adding a shed or gable dormer.

Shed Dormer

Shed dormers can be made any width, and they sometimes extend across the entire length of the house. They are considered by some to be less attractive than gable dormers, so they are usually placed at the back of the house. The sides, which should coincide with existing rafters if possible, are framed in the same manner as the sides of gable dormers. The low-slope roof has rafters framing directly into the ridge board (Fig. 141). Ceiling joists bear on the outer wall of the dormer, with the opposite ends of the joists nailed to the main roof rafters. The low slope of the dormer roof means that requirements for applying the roofing may be different from those for the main roof. Sizes and exposures for wood shingles should conform to those shown in Table 9; exposure for asphalt shingles are given in "Roof Coverings" in the "Rehabilitation" section. Roll roofing is generally used on low-pitched roofs.

Gable Dormer

If light and ventilation are the main deficiencies rather than additional space, gable dormers are often used. They are more attractive on the exterior than are shed dormers. However, because of the roof slope, gable dormers are usually limited to a small size. They are also more complicated to build than shed dormers.

The roof of the gable dormer usually has the same pitch as the main roof of the house. If possible, the dormer should be located so that both sides are adjacent to an existing rafter (Fig. 142). The rafters are then doubled to provide support for the dormer framing. The valley rafter is tied to the framing of the main roof by a header. After the window opening is formed, interior and exterior covering materials may be applied.

A critical matter in dormer construction is proper flashing where the dormer walls intersect the roof of the house (Fig. 112). Shingle step-flashing should be used at this junction. Step-flashing consists of aluminum or galvanized metal shingles bent at a 90° angle to extend up the wall

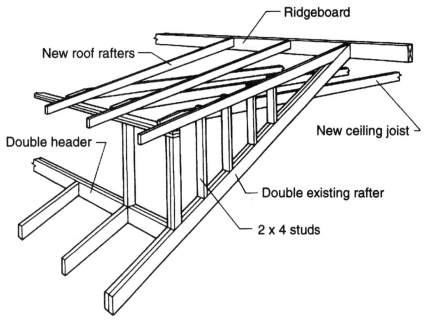

Figure 141—Shed dormer framing.

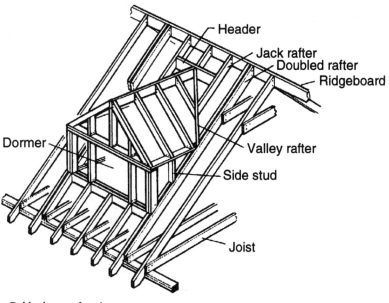

Figure 142—Gable dormer framing.

sheathing a minimum of 4 in. Use one piece of flashing at each shingle course, lapping successive pieces in the same manner as shingles. Apply siding over the flashing, allowing about a 2-in. space between the bottom edge of the siding and the roof. The cut ends of the siding should be treated with a water-repellent preservative.

Other
Considerations

Knee Walls

Side walls of the attic rooms are usually constructed by nailing 2- by 4- or 2- by 3-in. studs to each rafter at a point where the stud will be at least 5 ft long (Fig. 143). Studs should rest on a soleplate in the same manner as in other partitions. Blocking between studs and rafters at the top of the knee wall provides a nailing surface for the wall finish.

Ceiling

Nail collar beams between opposite rafters to serve as ceiling framing (Fig. 144). These beams should be at least 7-1/2 ft above the floor.

An alternate way to install the ceiling is to apply the ceiling finish directly to the rafters (Fig. 144). This type of ceiling is commonly called a cathedral ceiling.

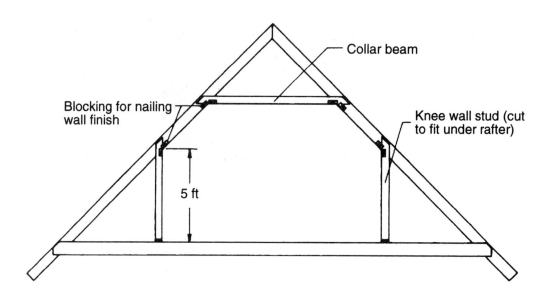

Collar beam

Blocking for nailing wall finish

Knee wall stud (cut to fit under rafter)

5 ft

Figure 143—Knee walls and blocking.

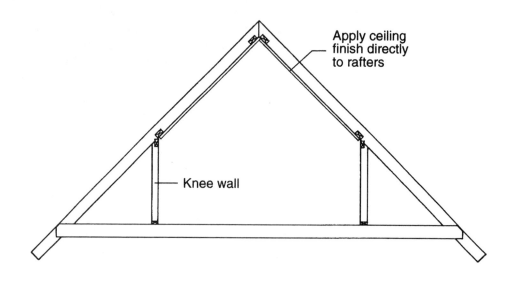

Apply ceiling finish directly to rafters

Knee wall

Figure 144—Attic finished with cathedral ceiling.

Chimney

If a chimney passes through the attic, it must be enclosed and the floor plan adjusted accordingly. Never frame into the chimney, and keep all framing members at least 2 in. from the chimney walls.

Horizontal Expansion

More space can be provided by constructing an addition to the house. As a preliminary step, local zoning and lot restrictions should be checked. General principles outlined in "Major Structural Features" and "Interior Structural Changes" in the "Rehabilitation" section should be reviewed.

Country houses can usually be expanded in any direction without restriction. In suburban developments, the minimum allowable distance from the front of the house to the street is usually uniform for all houses on a particular street; expansion to the front may not be possible. Required minimum setbacks on the sides may also leave no room for side expansion. In such cases, expansion to the rear is the only alternative.

When new space is created by additions, existing space can sometimes be converted to other uses as part of the overall plan. If another bedroom is needed but a larger living room is also desirable, perhaps the present living room can be converted to a bedroom and a large living room can be added. If it is more important to have a large modern kitchen, it can be added on and the old kitchen converted to a utility room, bathroom, or other living or work space.

The general style of the addition should be consistent with that of the existing house. Rooflines, siding, and windows should match the original structure as closely as possible.

Give attention to the creation of a coherent floor plan that will link the addition to the house. In some instances, it may be appropriate to use the satellite approach as described and illustrated in "Additions" in the "Condition Assessment" section.

Horizontal expansion of existing homes involves using standard techniques for new construction. Other sections of this handbook describe details for foundations, framing, floors, and other aspects to be considered when rehabilitating a house. Most of these instructions apply equally to new construction and to additions. The main requirement that is peculiar to the horizontal expansion of an existing house is making the connection between the addition and the house.

If the addition extends the length of the building, the structure, siding, and roofing must match the existing portion. To accomplish this, a complete roofing job may be necessary. If existing shingles can be matched, some shingles near the end must be removed to work the new shingles into the existing pattern. Short pieces of lap siding must also be removed so the new siding can be worked into the existing siding with end joints offset rather than as one continuous, vertical joint.

If the addition is perpendicular to the house, the siding can either match or contrast with the existing siding. However, the roofing material should match. Siding at the intersection of walls is applied with a corner board, as shown in Figure 110.

Kitchen Remodeling

New appliances and present concepts of convenience have revolutionized the kitchen in recent years. More space is required for the numerous appliances now considered necessary.

In spite of changing requirements, some principles of kitchen planning have applied for several years. The basic movements in food preparation are from the refrigerator to the sink, and then to the range. The four generally recognized arrangements for kitchens are the "U," the "L," the corridor, and the side wall (Fig. 24). The arrangement selected depends on the amount of space, the shape of the space, and the location of doors. If the kitchen is going to be a new addition, select the layout preferred, and plan the addition accordingly. The work triangle is smallest in the U and corridor layouts. The side-wall arrangement is preferred if space is quite limited, and the L arrangement is used in a relatively square kitchen plan in which table space is desired.

If kitchen cabinet space is adequate and well arranged, updating the cabinets may be the only desirable change. New doors and drawer fronts can be added to the old cabinet framing. Even refinishing or painting the old cabinets and adding new hardware can do much to improve an old kitchen. If a kitchen has adequate cabinets, the single improvement that will most update it is new countertops. Tops should be fabricated and installed by a good counter shop. Plastic laminated over particleboard or plywood backing is commonly used.

Doorways should be located to avoid traffic through the work triangle. Generally, doorways in corners should be avoided, and door swings should not conflict with the use of appliances, cabinets, or other doors. If swinging the door out would put it in the path of travel in a hall or other activity area, consider using a sliding or folding door. Sliding doors and their installation are expensive but may be worth the expense in certain situations.

Windows should be adequate to make the kitchen a light, cheerful place. The current trend toward indoor–outdoor living has fostered the patio kitchen, with large windows over a counter that extends to the outside to provide an outdoor eating area. This design is particularly useful in warm climates, but it is also convenient for summer use in any climate.

Where the sink is placed in relation to windows is a matter of personal preference. Many people like to look out the window while they are at the sink, but installing the sink along an interior plumbing partition is usually less costly than on an outside wall.

If the kitchen is quite large, it may be convenient to use part of it as a family room. The combined kitchen–family room concept can also be met by removing a partition to expand the kitchen or by adding on a large room. One method of arranging workspace conveniently in such a room is to use an island counter (Fig. 25), which can also serve as an eating counter for informal dining.

Much research in kitchen planning has been done by the USDA Agricultural Research Service and by state universities. Contact the local Agriculture Extension Service for information on kitchen planning from these sources. Many companies that build kitchen cabinets will assist in planning.

Heating, Air-Conditioning, Plumbing, and Electrical Systems

Updating or extending the heating, air-conditioning, plumbing, and electrical systems usually requires the services of skilled, licensed professionals. A full description of this kind of work is beyond the scope of this handbook. This chapter discusses construction considerations that are involved in accommodating these utilities.

Because heating ducts, plumbing stacks and drains, water piping, and electrical conduits must be run throughout the house, it is usually difficult to avoid cutting into structural members to accommodate them. Cut members can often be reinforced, and new walls or other framing can be built to accommodate these systems. If there is doubt about the critical nature of the member that must be cut and the nature of the reinforcement that may be needed, seek professional advice.

Cutting Floor Joists

Floor joists often must be cut or drilled to accommodate plumbing or electrical wiring. Cutting takes the form of notches at the top or bottom of the joist. Joists should be notched only in the

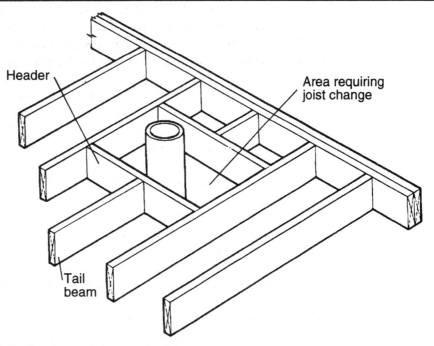

Figure 145—Framing to eliminate cutting joist.

end one-third of the span and to not more than one-sixth of the joist depth. If more alteration is required, the floor framing should be altered. Use headers and tail joists as shown in Figure 145 to eliminate the joist at that point.

Holes can be drilled through joists if the size of hole is limited to one-third of the depth of the joist in diameter and the edges of the hole are at least 2 in. from the top and bottom of the joist (Fig. 74). If a joist must be cut and the above conditions cannot be met, add a joist next to the cut joist, or reinforce the cut joist by nailing scabs to each side.

Utility Walls

Walls containing plumbing stacks or vents may require special framing. If a thicker wall is needed, it is sometimes constructed with 2- by 6-in. top and bottom plates and 2- by 4-in. studs placed flatwise at the edge of the plates (Fig. 146a). This framing technique leaves the center of the wall opening for running both supply and drain pipes through the wall.

Three-inch vent stacks fit into 2- by 4-in. stud walls; however, the hole for the vent requires cutting away most of the top plate. Scabs cut from 2 by 4s may be nailed to the plate on each side of the vent to reinforce the top plate (Fig. 146b). Also consider reducing the diameter of the vent stack in accordance with local regulations.

Fireplaces, Woodstoves, and Chimneys

From the standpoint of heat efficiency, which is estimated to be roughly 10 percent, conventional masonry fireplaces could be considered a luxury. However, if they are desired, their efficiency can be improved by installing factory-made circulating units. For even greater efficiency, use a fireplace insert or woodstove.

Conventional Masonry Fireplaces

Satisfactory performance by conventional masonry fireplaces can be achieved by following several general rules relating to the fireplace opening size, to flue area, and to certain other measurements. First, it is generally recommended that the depth of the fireplace should be about two-thirds the height of the opening. Thus, a 30-in.-high fireplace should be about 20 in. deep from the face to the rear of the opening.

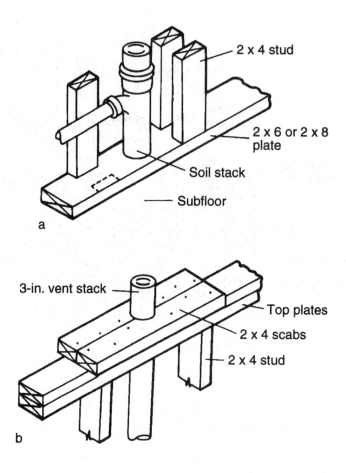

Figure 146—Plumbing in partitions: (a) thick wall for 4-in. soil stack, (b) reinforcing scabs for 3-in. vent stack in 2 by 4 wall.

The flue area (inside length times inside width) should be at least one-eighth of the area of the fireplace opening (width times height) when the chimney is 15 ft or less in height; the height is measured from the fireplace throat (Fig. 147) to the top of the chimney. If the chimney is more than 15 ft high, the flue area should be at least one-tenth of the area of the fireplace opening.

Thus, a fireplace that is 30 in. wide and 24 in. high (720 in²) would require a minimum of an 8- by 9-in. rectangular flue (or a 10-in.-diameter circular flue) when the chimney height is 15 ft or more.

A steel angle iron should be used to support the brick masonry over the fireplace opening. The bottom of the inner hearth, the sides, and the back should be built of heat-resistant firebrick. The outer hearth should extend at least 16 in. out from the face of the fireplace to provide protection against flying sparks; it should be made of noncombustible materials such as slate or tile. Other details relating to clearance, framing of the wall, cleanout opening, and ash dump are shown in Figure 147.

An adjustable damper should be used in the throat to control the opening. The smoke shelf (top of the throat) helps to prevent back drafts and should be about 8 in. above the top of the fireplace opening (Fig. 147). The shelf is concave to retain any rain or melted snow that comes down the chimney.

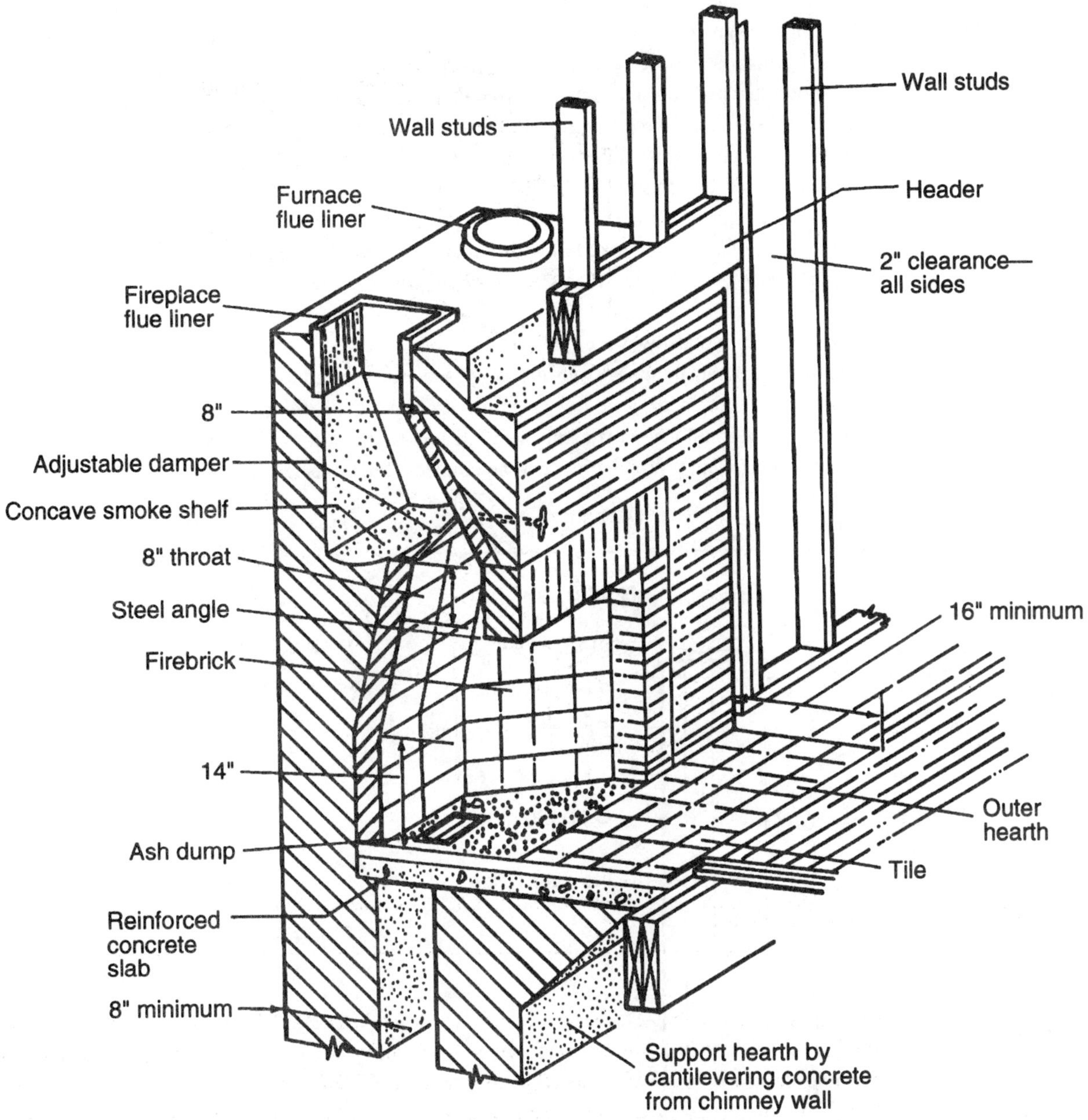
Wall studs
Wall studs
Header
Furnace
flue liner
2" clearance—
all sides
Fireplace
flue liner
8"
Adjustable damper
Concave smoke shelf
8" throat
Steel angle
16" minimum
Firebrick
14"
Outer
hearth
Ash dump
Tile
Reinforced
concrete
slab
8" minimum
Support hearth by
cantilevering concrete
from chimney wall

Figure 147—Masonry fireplace components.

Air-Circulative Fireplace Forms

The heating capacity of a fireplace can be increased by using a factory-built, steel air-circulating form, which usually includes the firebox sides and back plus the throat, damper, smoke shelf, and smoke chamber. The sides and back of the circulator are double-walled, enclosing a space within which air is heated. Cool air is introduced into this space near the floor level and, when heated, rises and returns to the room through registers located at a higher level. Air-circulating forms also help to assure satisfactory performance of the fireplace because their dimensions have been engineered for optimum performance.

The fireplace form is usually set on a firebrick floor laid on a reinforced concrete base. The chimney flue is constructed of conventional masonry, beginning at the top of the form. The form is enclosed in masonry, with decorative masonry facing the room. Small fans are often installed to increase the heating efficiency of the unit.

Prefabricated Fireplace Units

Factory-built fireplace units include all fireplace and chimney components from the hearth to the chimney cap. These units are called zero-clearance units because they can be installed on wood floors and against wood framing (Fig. 148). The units have steel walls and include insulation that protects adjacent wood members from excessive heat. They frequently include screens, glass doors, circulating fans, and allowances for external air-supply ducts. Zero-clearance fireplace units can be placed on the floor or on a raised hearth constructed of wood; they can be faced with stone or brick for a traditional appearance or enclosed with gypsum board and trimmed in wood. The insulated steel chimney pipe can be extended up through the interior of the house and boxed in with wood framing or housed in a wood-framed "chimney" on the exterior, covered with the same style of siding as the house.

Woodstoves

Some types of woodstoves are free-standing; others are designed to be inserted into fireplace openings. Their air intake is controlled to produce slower, more efficient combustion than is possible with fireplaces, with less loss of room air up the chimney. Some airtight woodstoves can provide a combustion efficiency as high as 30 percent (a typical older gas furnace is about 60 percent efficient). Woodstoves can be connected to either insulated steel or masonry flues (Fig. 149).

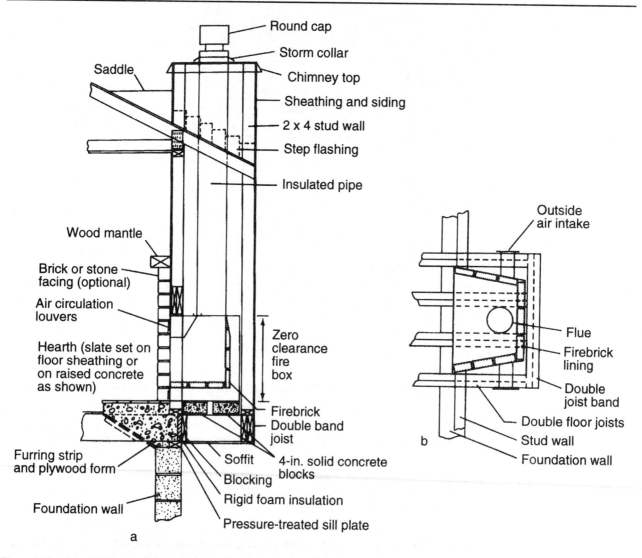

Figure 148—Zero-clearance fireplace in external wood chimney: (a) side view, (b) top view.

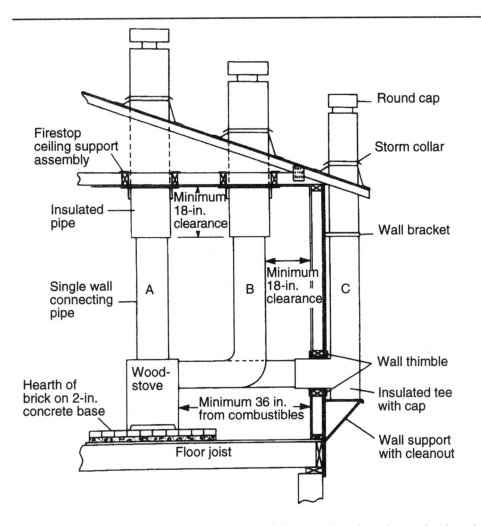

Figure 149—Woodstove: (A) top-mounted flue, (B) rear flue through ceiling, (C) rear flue through wall.

Some stoves are made with a single wall of steel or cast iron, which radiates heat. Others are enclosed in thin steel jackets, allowing air to circulate between the stove and the jacket. The cooler outer jacket is a desirable safety feature. Air enters and leaves the space between the stove and jacket through vents, and heats the room by convection. Fans are sometimes used to improve circulation. In some systems, the heated air may be distributed to other rooms through ducts. Some models contain coils through which water circulates; the heated water can be employed for domestic uses or for space heating. Water can be pumped to various locations or circulated by convection.

As woodstoves have increased in popularity, failure to observe proper fire-safety precautions in construction and installation has resulted in fires and accidents. It is especially important to remember that wood and other combustibles can be heated to the flash point even if there is no direct contact between the hot stove and the combustible material. Sufficient heat to ignite combustibles can be passed from the stove across airspaces by convection and radiation, or through intervening noncombustible materials, such as masonry, that are in contact with the combustible material. The following precautions should be taken whether or not they are specified in instructions provided by the manufacturer:

1. When an uninsulated metal pipe or thimble passes through a wall, ceiling, or other framing, an airspace of at least 6 in. should be maintained around the pipe, with fiberglass insulation inserted between the pipe and the structure at all points.

2. Clay thimbles should not be run directly through concrete block or other nonflammable masonry. The thimble, masonry, or both, may crack, allowing heat to rise within the masonry cavities and ignite the wood sill. An insulated steel pipe should be used in this case, or a steel pipe can be passed through the thimble with an airspace maintained between the pipe and thimble.

3. Stove manufacturers specify minimum safe distances from walls and other combustible materials. The Consumer Products Safety Commission states, "Be sure to follow the manufacturer's instructions carefully and be certain that the equipment has been approved by a nationally recognized testing agency. The stove should be kept at least 3 feet from walls, ceiling, and furnishings unless instructions indicate otherwise." Requirements for woodstoves are covered in Chapter 11, "Equipment, General," of the Council of American Building Officials (CABO) One- and Two-Family Dwelling Code. The code specifies that stoves be placed 4 ft from walls and ceilings unless these surfaces are covered by specified types of protective materials, in which case the requirement is reduced to 3 ft. Details are provided in tables M–1102.1 and M–1102.2 in the 1989 edition of the code.

4. Free-standing woodstoves should be set on brick or concrete hearths. Bricks of standard 2-3/4-in. thickness or 3 in. of concrete should be used.

Masonry Chimneys

Chimneys can be constructed of masonry units supported on a suitable foundation or of insulated stainless steel pipe. They must be structurally safe and capable of producing sufficient draft for fireplaces, stoves, and/or other fuel-burning equipment.

A masonry chimney should be built on a concrete footing of sufficient area, depth, and strength for the imposed load. The chimney footing should be below the frostline. A cleanout door is usually included in the bottom of the chimney below the thimble, if one is present. Masonry chimneys are usually free-standing and constructed in such a way that they neither support nor are supported by the house structure.

Masonry chimneys should be separated from woodframing, subfloor, and other combustible materials by at least 2 in. This space should be fire-stopped at each floor with a noncombustible material, such as fiberglass insulation (Fig. 150). Subfloor, roof sheathing, and wall sheathing should have at least a 3/4-in. clearance from the chimney.

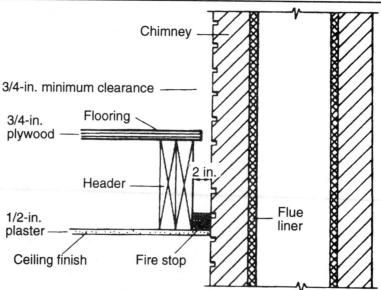

Figure 150—Chimney clearances for wood-frame construction.

The size of the chimney depends on the size and number of flues. Each fireplace, stove, or furnace should have a separate flue. Flues should be separated by 4 in. of masonry (Fig. 20). A concrete cap is usually poured over the top course of brick (Fig. 19). Precast or stone caps are also used. The height of a chimney above the roofline usually depends on its location in relation to the roof ridge. If the chimney is within 10 ft of the roof ridge, it must extend a minimum of 24 in. above the ridge. If the chimney is more than 10 ft from the roof ridge, the top of the chimney must extend a minimum of 24 in. above the highest point on the roof that is within 10 ft of the chimney. In addition, the chimney must extend at least 36 in. above the highest point on the roof immediately adjacent to the chimney (Fig. 21). Flashing for chimneys is illustrated in Figure 151.

Round or rectangular fire-clay flue linings are normally used for masonry chimneys. Rectangular flue lining is made in 2-ft lengths and in various sizes from 8 by 8 in. to 24 by 24 in. Wall thicknesses vary with the size of the flue; smaller linings have a 3/4-in.-thick wall. Round tiles 8 in. in diameter are commonly used for the flues of heating equipment, although larger sizes are also available.

Flue lining should begin at least 8 in. below the thimble for a connecting smokepipe or vent pipe from the stove or furnace. For fireplaces, the flue liner should start at the top of the throat and extend to the top of the chimney. Flue liners should be installed well before the brick or masonry work, as it is carried up, so that careful bedding of the mortar will result in a tight, smooth joint.

Standard concrete flue blocks are available for building less expensive chimneys. These blocks are 8 in. high by 16 in. square or larger, with holes in the center sized to fit standard flue liners. Other blocks have half-circular holes on one side; two of these together form a circular opening through which a thimble can be placed for a connecting smokepipe or vent.

Insulated Steel Chimneys

Insulated steel chimneys are made in sections 12 to 36 in. long that are joined together to form a long pipe. One type has a triple wall consisting of three pipes with spaces between them through which air circulates to remove heat and cool the pipes. The inner pipe is made of stainless steel; the outer pipes are galvanized. Another type consists of double-wall stainless-steel pipe with insulation between the walls.

Both types come with a full line of accessories including tees, wall supports and brackets, roof supports and flashing, storm collars, caps to keep rain from going down the flue, and spark arresters. Either type can be fully exposed to weather, extended up through the interior of the building, or enclosed in a wood chimney.

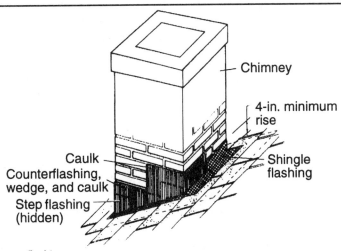

Figure 151—Chimney flashing.

Steel chimneys are not usually damaged by flue fires, which can crack clay flues. If creosote buildup is ignited in a steel flue, the fire will continue until the creosote burns off and, if the manufacturer's installation recommendations have been followed, the flue may not be damaged. However, it is a good idea to clean the chimney periodically to prevent chimney fires. Steel flue pipe should bear a label signifying approval by Underwriters Laboratories, Inc. or other recognized certifying agency.

Garages and Carports

A garage can be attached or detached. A carport is a roofed, open structure for sheltering vehicles. The attached garage has several advantages: it can enhance the architectural lines of the house, it is warmer during cold weather, and it provides convenient storage space. It also protects people who enter or leave vehicles, and it provides a short, direct entrance to the house. An attached garage is also less expensive to build than a detached garage because it shares a wall with the house.

If there is considerable slope to a lot, a basement garage may be a possibility. A basement garage generally costs less because it does not require an additional structure.

Carports are usually attached to the house. Because of its open walls, a structurally independent, detached carport is difficult to build. Storage units often are built along the open side or at the end of a carport, and can provide partial screening.

Addition of Garages or Carports

Larger cars are up to 18 ft long. Although the garage need not necessarily be designed to take the largest cars, it is good practice to provide a minimum distance of 19 to 21 ft between the inside faces of the door and the opposite wall. If additional storage or work space is desired, more depth is needed. The inside width of a single garage should never be less than 11 ft. The recommended minimum outside size for a single garage, therefore, would be 12 by 20 ft. A double garage should be not less than 20 by 20 ft in outside dimensions to provide reasonable clearance. Adding a shop or storage area would increase these dimensions.

The foundation wall for an attached garage should extend below the frostline and about 8 in. above the exterior final grade level. It should be at least 6 in. thick. The sill plate should be anchored to the top of the foundation wall with anchor bolts spaced about 8 ft apart, with at least two bolts in each sill piece. Extra anchors may be required at the sides of the main door.

If fill is required below the floor, it should be sand or gravel. If some other type of soil fill is used, it should be well-compacted. If these precautions are not taken, the concrete floor is likely to settle and crack.

The concrete floor should not be less than 4 in. thick. It should be laid with a pitch of about 2 in. from the back to the front of the garage. Welded wire mesh is often used to help control cracks; however, it should be placed in the top one-third of the concrete. The garage floor should be set about 1 in. above the drive or apron level. It is desirable to have an expansion joint between the garage floor and the driveway or apron.

The framing of the side walls and roof and the application of the exterior covering material are similar to that of the house. Interior studs can be left exposed or covered with some type of sheet material. Building codes require that the wall between the house and the attached garage be covered with fire-resistant material such as 1/2- or 5/8-in. Type X gypsum board. Consult local building regulations and fire codes before construction begins.

Garage Doors

The most commonly used garage doors are the overhead sectional type (Fig. 152). They are made in four or five horizontal hinged sections and have a track extending along the sides of the door opening and under the ceiling with a roller for the sides of each section. They are

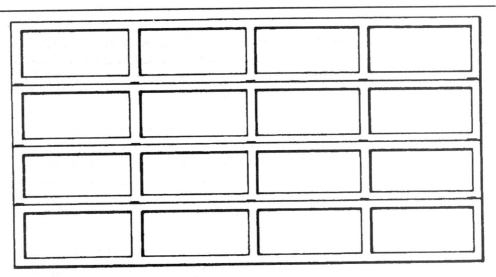

Figure 152—Sectional overhead garage door.

opened by lifting, and are easily adapted to automatic electric openers with remote-control devices. The standard size for a single door is 8 or 9 ft wide by 6-1/2 or 7 ft high. Doors for two-car-wide garages are usually 15 or 16 ft wide.

Doors vary in design, but those most often used are the panel type with wood stiles, rails, and panel fillers; a glazed panel section is often included. Translucent fiberglass and embossed steel or aluminum doors are also available. Clearance from the top of the door to the ceiling is usually about 12 in., although low-headroom brackets are available that can reduce the required clearance to 6 in.

The header beam over garage doors should be designed for the dead load and live load that might be imposed by the roof above, or a sagging header may result. If this header also carries a floor, the floor loads must be considered. Three 2- by 12-in. members, 18 ft long, are typically required for built-up wood beams over 16-ft door openings; however, in regions that get heavy snows, larger beams may be required. Garage doors in trussed-gable end-walls do not carry roof loads and therefore do not require a structural header.

To keep the garage warmer in cold climates, some overhead door units can be ordered with insulation and with weatherstripping for the perimeter of the door. Weatherstripping is typically made of vinyl for head and side jambs and for contact with the floor.

Carports

Carports are often built with 4- by 4-in. solid wood posts at all corners and at other intermediate points determined by the size or span of the load-bearing headers. The posts should be pressure-treated with a wood preservative. With the typical 10- to 12-ft spacing between posts, two or three 2- by 12-in. headers may be used to span between the posts.

Metal post bases are often used to fasten posts to the concrete slab. The load-bearing header is either bolted or nailed to the posts. To ensure that it remains securely upright, the header may be set over a pin, such as a short section of 1/2-in. reinforcing rod set in a hole in the top of the post and in the header. The connectors must be able to resist strong wind-uplift forces. Clearances for automobiles should be the same as for garages, to allow for the possibility that the carport may be closed in at a later date.

Conversion of Garage to Living Space

A garage that was originally built integrally with the house can be converted to living space by insulating it and adding floor, wall, and ceiling finish. In addition, windows and/or heating may be required.

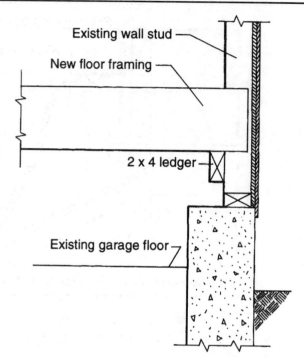

Figure 153—Framing to bring new garage floor to the level of the house floor.

Floors

The new floor may be applied directly over the existing concrete slab or, if headroom is sufficient, over new floor framing suspended above the slab. If the ceiling of the finished garage is at the same elevation as the house ceiling, new floor framing can be installed at the same level as the house floor. The framing may rest on the foundation wall or be supported on ledgers nailed to the wall studs (Fig. 153). Consult Table 12 to determine the correct size for the required span. The floor is constructed in the same manner as a conventional floor, as described in "Floor Systems" in the "Rehabilitation" section, and is insulated in the same manner as a crawl space, described in "Types of Insulation" in the "Condition Assessment" section. The garage roof is often lower than the house roof, so the new floor must be placed directly on the concrete slab one or two steps below the house floor. Finish flooring can be installed with the same materials and techniques as for a basement slab. In all but very warm climates, insulation should be provided. If the new floor is built on sleepers over the concrete, insulation may be placed between the sleepers.

Another method of insulating is to dig the soil away from the foundation and apply rigid insulation to the outer face of the foundation wall to the depth of the footing (Fig. 154). The insulation must be moistureproof, such as extruded polystyrene, and can be attached with a mastic recommended by the insulation manufacturer. The insulation should be covered above grade with a material suitable for contact with the ground, such as cement board or preservative-treated plywood. If subterranean termites are a threat, the soil should be poisoned during the backfill operation.

Walls

Garage walls are not usually insulated, so blanket insulation with a vapor retarder on the inside face should be installed in the space between studs. Then any type of wall finish may be applied. Half-inch-thick gypsum board or wood paneling is typical.

Frame any additional doors and windows following the instructions given in "Window and Door Openings" in the "Rehabilitation" section. It may be convenient to use the existing garage door opening for a sliding glass door or a window wall. If not, it may be filled in with conventional 2- by 4-in. framing and covering materials.

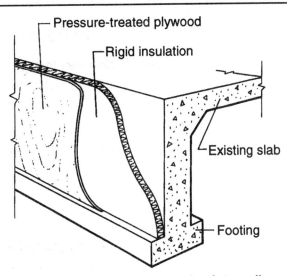

Figure 154—Application of insulation to outer face of garage foundation wall.

Ceiling

The ceiling can be installed using standard methods and materials employed in new construction. Half-inch-thick gypsum board is typical. Be sure to insulate it well and to ventilate the attic space.

Deck, Patio, and Porch Additions

Decks, patios, and porches are all means to economically extend a home beyond its totally enclosed portions and can be thought of as outdoor extensions. Both open and enclosed porches were common on Victorian-style farm and ranch homes, for example. They provided space for sitting and relaxing and also for storage. Patios are also an extension of the indoors to the outdoors. They are not generally attached to the house and are usually built of masonry material set level with the ground.

The popularity of wood decks has increased greatly. They can expand or frame a view to increase the homeowner's enjoyment of the outdoors. On level ground, they can be built as an attachment to the house or they can be detached. On some steep hillsides, decks are the only means for extending the living quarters. Because their construction is generally not complicated, decks also provide a means for many homeowners to exercise their own skills in construction and home improvement.

However, decks, patios, and most porches are built in such a way that at least some members are exposed to the weather. As a result, certain precautions must be taken to ensure that the structure will have an acceptable service life.

Planning and Design

A first step in planning an outdoor extension is to determine the requirements and limitations of the local building code. Floor load requirements, construction requirements concerning the minimum height and opening size of railings to prevent children from falling through, and other aspects of porch and deck construction vary by locality and must be determined before a deck is designed. Other considerations include view, orientation with respect to the prevailing winds, morning and evening shade or sun, steepness of the site, and desires of the owner.

It can sometimes be difficult to add an outdoor extension onto an existing structure. It may require creating a new doorway, adding a sliding door, or devising a walkway to get from one part of the house to the outdoor living area. If the outdoor extension is to have a roof, make certain that it can be properly tied into the existing structure and that an acceptable slope can be maintained. If the roof will have a minimal slope, continuously sealed roofing or built-up roofing may be required instead of shingles.

Because of the great interest in outdoor extensions, numerous detailed publications are available from trade associations, retail lumberyards, and other sources.

Material Selection

Because of its relatively low cost, workability with common tools, and ease of transport, wood is often the preferred material for outdoor extensions. Wood products on the market offer a number of choices in regard to size (lumber compared to plywood), durability, strength, and other factors. Masonry products are also used for deck and patio construction. Concrete can be finished smooth or with different textures and design. Brick is a popular construction material for exterior walls, particularly in the south where freezing, and thus heaving and cracking, is not a problem. Bricks, of course, come in a variety of colors and can be laid in a number of different patterns.

Exposed lumber should be pressure-treated with an acceptable preservative or, if not treated, should be from the heartwood of a naturally durable species. Naturally durable heartwood of redwood, cedar, and cypress traditionally have been used if moisture, decay, and insects are potential problems. Today, much lumber produced from these species is from smaller trees and can contain substantial sapwood. The lighter-colored sapwood is not durable and should not be used in unprotected areas. Also, most lumber produced from these species commands a premium price and is used for aesthetic applications. Therefore, most lumber used for outdoor construction is pressure-treated with a wood preservative. Pressure-treated wood that will be in contact with soil should be marked "for soil contact" or the quality mark should show a double-digit number such as LP22. Selection of this material for different applications and for quality is discussed in "Decay From Moisture" in the "Rehabilitation" section. Remember, material in contact with the ground will require a higher retention of the preservative than will material that is not in ground contact. Also, if the material is placed in salt or brackish water where marine borers may be present, special treatments and penetrations as prescribed by the American Wood-Preservers' Association will be required.

Plywood is often used if a solid deck covering is desired. Only exterior-grade plywood marked C–C plugged exterior or underlayment exterior (C–C plugged) or higher grades of A–C or B–C should be used. The species and thickness needed will depend on the joist spacing and design loads. Specialty plywood, such as that with a high-density overlay (HDO) and a hard, phenolic resin impregnated fiber surface, which gives a screened, skid-resistant finish often used for boat decks, may be used for deck covering; medium-density overlay (MDO), which has a softer resin-fiber surface, may also be used. These plywood surfaces require either a high-performance deck paint or an elastomeric deck-coating system. Unless the plywood can be protected from moisture, it should also be pressure-treated with a paintable wood preservative. Most specialty-type plywoods need to be ordered by a retail lumberyard. Therefore, it is best to check with a knowledgeable individual concerning availability, cost, and suitability of such plywood before making a commitment to its use.

Structural Considerations

Outdoor extensions such as elevated decks and porches are subject to loads ranging from only one or two people to being filled to standing-room-only conditions. Therefore, they should be thought of as structures, and appropriate engineering design information for the materials and application should be applied to their design. Standard plans are available from building material suppliers for some deck types.

The allowable spans for decking, joists, and beams and the size of posts depend not only on the size, grade, and spacing of the members but also on the species. Species such as Douglas Fir, Southern Pine, and western larch allow greater spans than do some of the less dense pines, cedars, and redwood. Normally, deck members are designed for about the same load as the floors in a dwelling. Before beginning construction, check the requirements of the local building codes or trade association recommendations for spans by species, grade, and anticipated design load.

The arrangement of the structural members can vary somewhat because of orientation of the deck, position of the house, slope of the lot, and so forth. However, basically, the beams are supported by the posts (anchored to footings) and they, in turn, support the floor joists (Fig. 155). The deck boards are then fastened to the joists. For 16-in. joist spacing, deck boards may be 2 by 4s or a special deck lumber that is a full 1 in. thick by 5-1/2 in. wide and with rounded corners. For porches, tongue-and-groove 1 by 4s or 1 by 6s are normally used to construct the deck surface.

Railings are located around the perimeter of the deck or porch, if required for safety. The main railing supports may be attached to the floor joists, or the posts used for support may extend through the floor and serve as attachments for the railing. Low-level decks may be constructed without railings. In high, free-standing decks, the use of diagonal cross-bracing from post to post is a good practice.

Porches and decks are normally attached to the main framing members (sills, studs, ceiling joists, roof rafters, and so forth) of the house. The siding is normally removed at the point of attachment so a firm joint can be constructed. Figures 156 a, b, and c show three ways to attach a deck to a house. The wood framing members of the house will probably not be pressure-treated nor of naturally durable species. Therefore, flashing must be installed wherever water can accumulate, especially between the house framing members and the deck. Applying a water-repellent preservative to the nontreated wood may also be helpful. To prevent water accumulation, the first deck board should be spaced about 1/2 in. from the house.

For low-level decks, the joists may be placed directly on footings. For high-level decks, 4 by 4 or 4 by 6 columns or posts of pressure-treated wood are generally used. These posts may be set directly in the ground or placed on concrete footings. If placed on concrete footings, the posts are normally supported just above the concrete with a post anchor. The bottoms of the footings or the bottoms of the posts should be placed below the frostline, which is generally 3 ft or more in northern climates.

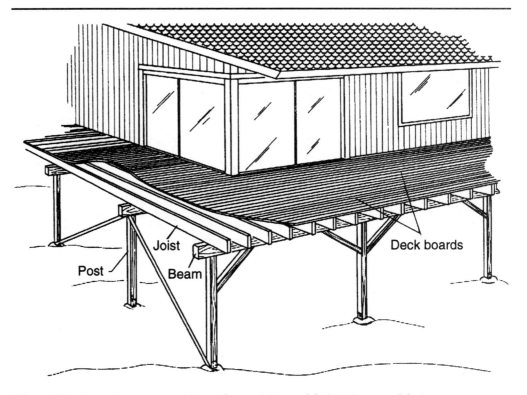

Figure 155—General arrangement of posts, beams, joists, and decking for a wood deck.

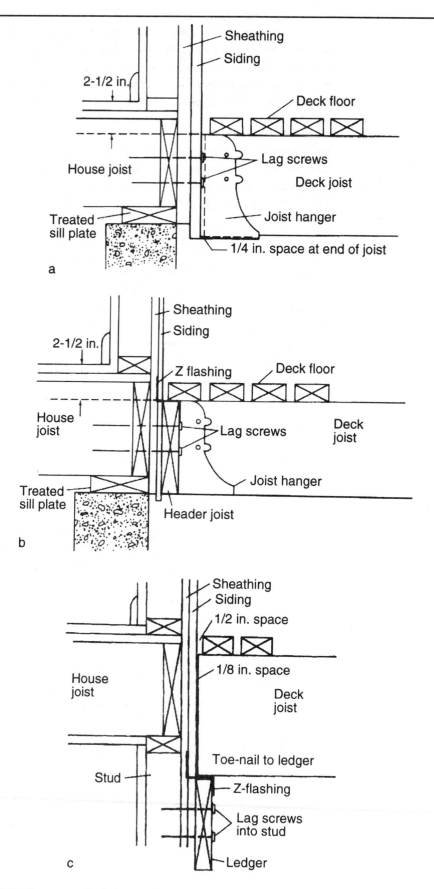

Figure 156—Common methods for attaching deck hangers (a), with header joist (b), and using ledger plate (c). (From *Construction Guide for Exposed Wood Decks*, Agriculture Handbook 432.)

Porches can be built on concrete footings or other masonry structures as described for decks. They can also be built on concrete slabs poured directly on the ground or onto a foundation wall. Masonry footings or foundations should be installed to allow the finished floor to have an outward slope of 1/8 in. per foot. If the area under a porch is to be closed or to have a foundation wall, be sure to allow for ventilation and access to inspect for termites, if appropriate. The soil should be chemically treated to prevent termite infestations and covered with a polyethylene sheet to prevent upward moisture movement. Insulation can also be added, if needed.

Fasteners

The strength and utility of any wood structure is dependent upon the fasteners used to hold it together. The most common fasteners are nails, followed by bolts, metal connectors and straps of various shapes, lag screws, and screws. An important factor for the outdoor use of fasteners is the finish. The most commonly used outdoor fasteners are hot-dip galvanized, aluminum, or stainless steel, which keep the wood from staining when it is exposed. Do not use standard ferrous fasteners on wood treated with the common salt-type wood preservatives because the fastener will corrode and deteriorate. A rusted nail, washer, or bolt head is not only unsightly but the connection will lose strength gradually as the corrosion continues.

The best assurance of withdrawal resistance is the use of annular grooved (ring shank) or spirally grooved nails (Fig. 157). The value of these nails is their capacity to retain withdrawal resistance even after repeated wetting and drying cycles. When nailing through two pieces placed wide face to wide face, the nails should go completely through both pieces far enough so that the ends can be easily clinged. If the nails are driven at about a 30° angle, they are less likely to loosen.

Bolts are one of the most rigid fasteners. They may be used for small connections such as railings-to-posts and for connecting larger members. Carriage bolts, or those with a round head, are commonly used for wood construction. A squared section just under the round head of the bolt resists turning as it is tightened. A washer should always be used under the nut. Bolt holes should be the exact diameter of the bolt. Avoid crushing wood under the head or washer and nut of a bolt.

Finishes

The flooring of decks and, to a lesser extent, porches is exposed to the elements and therefore requires special finishing. Porch flooring can be painted with the same techniques as described for siding with one exception: the top coat should be exterior porch enamel, which is specially formulated to resist abrasion and wear.

Because most parts of a deck are flat, horizontal surfaces that are fully exposed, finishes that form surface films are short-lived. A water-repellent preservative or a semitransparent penetrating oil-based stain is generally the most effective finish. Because of the wear and exposure, water-repellent preservatives may need to be renewed yearly; penetrating stains will usually last 2 to 3 years. The service life of light-colored stains is usually shorter than the dark ones

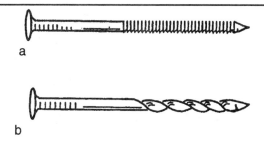

Figure 157—Annular grooved or ring shank (a) and spirally grooved (b) nails provide greater holding strength and less loosening. (From *Construction Guide for Exposed Wood Decks,* Agriculture Handbook 432.)

because the light-colored ones have less pigment to protect the wood. The application of these finishes is described in "Application" in the "Rehabilitation" section. These penetrating finishes are easily renewed. Paints and solid-color stains tend to peel when they are fully exposed, and they are more difficult to restore. They are not recommended for decks.

Decks should be constructed of wood that is pressure-treated with waterborne preservatives to prevent decay or of a naturally durable wood (common names of waterborne preservatives are CCA, Wolman, or Osmose). Therefore, decks are often allowed to weather to a natural gray. The bright color of the wood can be restored by applying commercial products (called deck cleaners, brighteners, or restorers). These products may remove the weathered wood surface, but some care should be exercised not to remove excess wood. Color can also be restored using liquid household bleach containing 5-percent sodium hypochlorite. The bleach is usually diluted with water (one part bleach, three parts water) before it is applied to the deck. Rinse the bleach solution off the deck with water. If the deck is to be finished after cleaning, allow 1 or 2 days drying time.

Maintenance

A home that is well designed and constructed or rehabilitated with the proper materials should have very few maintenance problems for the first 10 years or so of its life. Nevertheless, the occupant or owner should routinely check all areas of the structure to make certain that something has not deteriorated.

General Considerations

The approximate life expectancy for major appliances and other equipment in most homes is listed in "Budgeting for Maintenance" in the Introduction. Normally when this equipment fails, it is replaced immediately because the home becomes nonfunctional.

Other problems may develop more slowly and therefore may go undetected or just be neglected. At least twice a year, make sure the gutters and roof valleys (if present) are free of any debris. This is best done in the fall after all leaves are off the trees and again in the spring after the trees have leafed out and flowering is complete.

Make certain that the downspouts are clean and that all water is draining away from the foundation. If it is not, moisture problems in the basement or crawl space are likely to occur. If the house has a basement and a drain tile around the footing, make certain that the tile is functional and not plugged by soil, roots, or other debris.

Check the basement and crawl space for any water or moisture accumulation. Be sure to check the sheathing under the commodes, tubs, and shower areas or where any plumbing may be leaking. If any dampness is present, make repairs immediately. Make certain that the vent areas of crawl spaces are open and free of any obstructions, including plants, and that the soil is dry or covered with a vapor retarder.

Observe the inside and outside of any perimeter walls for evidence of termite infestation. Shelter tubes usually indicate an infestation unless the insects have limited their pathways to the inside of hollow-core blocks.

Observe the roof and roof line. If asphalt shingles are beginning to curl and break off and if significant amounts of shingle granules are found where water drains from the roof, new shingles may be needed. Wood shingles gradually erode or decay away until they are too thin to cover the roof adequately. Wood shingles also curl and crack with age. Examine the edge of the roof to make certain that water is not rotting the ends of the rafters, fascia boards, or sheathing.

Exterior Wood Finishing

The finish on exterior wood or wood-based surfaces requires the most attention. These surfaces are exposed to the weather and, as such, will need periodic refinishing or treatment.

Walls

The application and refinishing of exterior walls is thoroughly discussed under "Exterior Finishes" in the "Rehabilitation" section. In general, a well-applied two-coat (primer and one top coat) paint system will last 4 to 5 years, whereas two top coats may last up to 10 years. A penetrating oil-based stain applied using a two-coat wet system may last up to 8 years on rough wood in certain exposures, whereas a water-repellent preservative will last only 1 to 2 years.

Paints normally fail by soiling and chalking and should be recoated as soon as any wood is exposed, preferably before that stage is reached. Stains are normally recoated as the bare wood is exposed and as their color fades. Water-repellent preservatives are recoated every year or two or when water splashed onto the surface fails to bead up and run off.

Decks and Porches

Decks and, to a lesser extent, porches consist of members in a horizontal or flat position and are exposed to the weather; therefore, their wood finishes are severely exposed. Deck and porch finishes are discussed in "Finishes" in the "Rehabilitation" section. For decks, the wood should be pressure-treated with waterborne preservatives or it should be naturally durable wood. Deck surfaces are best treated with a water-repellent preservative, which may need annual renewal, or with a semitransparent oil-based penetrating stain, which may need renewal every 1 to 2 years. The advantage of these finishes is that they are easily renewed. Paint and solid-color stains will probably peel and need to be completely removed before being refinished.

Porch enamel can be used on a floor that is protected by a roof. Be sure to apply a water-repellent preservative and follow good painting procedures.

Maintenance of Wood Roofs

Wood roofs are subject to decay, so any practices that encourage quick drying after wetting should be followed.

First, leaves and other debris that often accumulate on roofs, particularly in valleys and gutters, trap moisture in the shingles, increasing the probability of decay. Therefore, loose debris should be routinely cleaned from roofs and gutters. Overhanging limbs and vines that produce excessive shade should also be removed.

Next, the roof should be checked for moss or lichen growth, which should be chemically treated. One simple method to prevent moss from developing on roofs is to use zinc, galvanized, or copper flashings. The normal erosion from these metals wil provide some control of moss (and mold and mildew) for about 10 to 15 ft downslope from the metal.

Brush treatment of selected chemicals can also provide some protection. A solution of copper naphthenate, copper-8-quinolinate, or other preservatives can be used to control moss, lichens, and surface decay and to prevent their growth for some time. Solutions are best applied by brush or by dipping. Follow manufacturers' recommendations for applications and safety carefully because wood preservatives can be toxic to plants, animals, and humans if used improperly. Humans, animals, and vegetation should be protected from drippings and runoff from the roof or gutters during application.

Certain commercial weed killers or herbicides can also be effective in moss control.

Regardless of the method used, surface treatments will not prevent serious decay problems within the shingle or in unexposed portions that are not treated. However, in some cases, they can help to lengthen the life of a wood roof by preventing the growth of moss and lichens, which retain moisture in the wood and thereby promote wood decay.

Caution

The pesticides—wood preservatives, mildewcides, and fungicides—described in this report were registered for the uses described at the time the report was prepared. Registrations of pesticides are under constant review by the Environmental Protection Agency. Therefore, consult a responsible state agency on the current status of any pesticide. Use only those that bear a federal registration number and carry directions for home and garden use.

Pesticides that are used improperly can be injurious to humans, animals, and plants. Follow the directions and heed all precautions on the label. Avoid inhaling vapors and sprays; wear protective clothing and equipment if these precautions are specified on the label.

If your hands become contaminated with a pesticide, do not eat, drink, or smoke until you have washed. If you swallow a pesticide or if it gets in your eyes, follow the first-aid treatment given on the label and get prompt medical attention. If a pesticide gets onto your skin or clothing, remove the clothing immediately and wash your skin thoroughly.

Store pesticides and finishes containing pesticides in their original containers out of the reach of children and pets, under lock and key. Follow recommendations for disposing of surplus finishing materials and containers. Scraps of chemically treated wood or finished wood should never be burned for heat or disposal because toxic fumes may be released. Dispose of the material through ordinary trash collection or bury it.

Appendix

MOST FREQUENT DANGER AREAS DUE TO TERMITES

3/ Concrete porches with earthfill underneath pose a special hazard. Often, wood framing members are in contact with the fill.

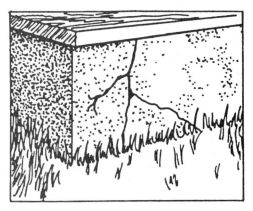

1/ Cracks in concrete foundations allow termites hidden access from soil to sill.

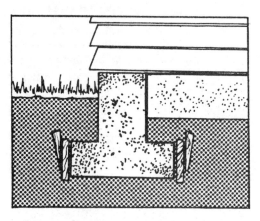

4/ Form boards left in place contribute to termite food supply.

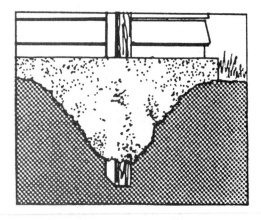

2/ Posts set in concrete may give a false sense of security. What if they are in contact with the soil underneath?

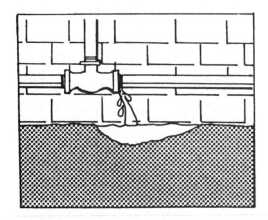

5/ Leaking pipes and dripping faucets keep the soil moist. Excess irrigation has the same effect. Downspouts should carry water away from the building.

6/ Shrubbery which blocks the air flow through vents causes the air underneath house to remain warm and moist — an ideal climate for termites!

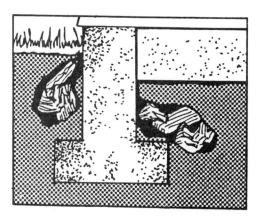

7/ Debris under house supports termite colony until population becomes large enough to attack the house itself.

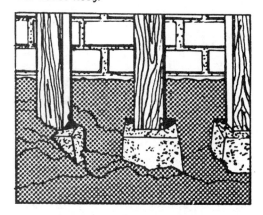

8/ Foundation walls or footings which are too low, permit wood to contact the soil.

9/ Stucco or brick veneer carried down over concrete foundation permits hidden entrance between exterior and foundation, if bond fails.

10/ Planters built against the foundation allow direct access to unprotected veneer, siding, or cracked stucco.

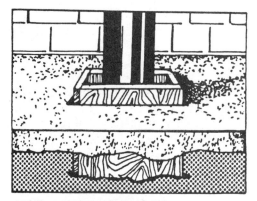

11/ Forms left in hole of slab where bathtub drain enters building provide a direct route to inner walls.

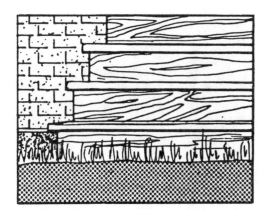

12/ Porch steps in contact with untreated soil literally offer termites a stairway to your home.

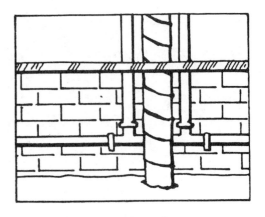

14/ Paper is made of wood. Don't leave paper collars around pipes.

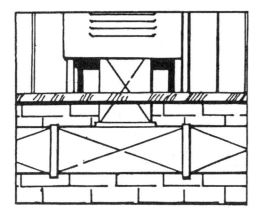

13/ Heating unit accelerates termite development by maintaining warm soil for colony on a year-round basis.

15/ In a house that was once termite-free, do-it-yourselfers can provide access to termites by building trellises and other adornments which provide a direct link from soil to wood.

PREVENTION IS THE BEST CURE

As already seen, most of the danger areas can be rendered harmless by any homeowner simply by moving soil as far away as possible from any wooden member of the house. But careful! Chances are that, before your house was built, the builder treated the soil underneath and adjacent to the slab or pier foundations with a termite toxicant. If done properly, the treatment will prevent termites from entering through or around the foundation for many years. In fact, research by the U. S. Forest Service has revealed four chemicals—aldrin, chlordane, dieldrin and heptachlor—which create a termite-proof barrier for at least 25 years. None have failed yet! So if you transport much soil from the immediate vicinty of foundation piers or a concrete slab foundation, the remaining soil should be chemically treated again.

If you find or suspect termites call a responsible pest control operator. As said before, don't panic. Take your time and get two or three cost estimates from established firms. Check references of the operator and beware of firms that:

• quote a price based on gallonage of material used(get estimate of total price for the job).

• profess to have a secret formula or ingredient for termite control. Chemicals tested by the U. S. Forest Service are the best known to man and are not expensive.

• have no listed phone number.

• show up without an invitation and use evidence of termites in trees as an excuse to inspect the house.

• also want to trim trees and do general repair work as part of the contract.

WHAT HOMEOWNERS ASK ABOUT TERMITES

Scientists at the Wood Products Insect Laboratory are among the nation's leading authorities on the biological habits of termites and methods of termite prevention and control. Here are their answers to some of the questions they receive from around the world:

How can I tell if the damage to my door jamb is caused by termites or decay?

Strip away the outer surface of the board. If termites have been there, you'll find that the softer springwood hasn't been eaten away while the harder summerwood is intact, giving a honeycomb effect. Termites also leave light brown specks of excrement and earth in their path. When decayed, the wood is soft to the touch and, of course, there's no earth or tunnels inside

I prefer not to use chemicals. Is there any biological control for termites?

Not yet. We and other scientists are searching for biological controls but none have been found to be very effective so far. However, all four recommended chemicals are registered for use against termites at this time. The U.S. Forest Service has shown that these chemicals have moved only a few inches laterally or downward through sandy loam soil after two decades of heavy rainfall. Thus, the risk of contaminating water resources is minimal. Moreover, most of the insecticide is under the building and not exposed to the environment. In the event that any or all of these insecticides become unavailable to pest control operators, the Forest Service is continuing its research effort to find effective substitutes.

I plan to build my house on a concrete slab. Isn't this the safest from termite attack?

Actually, it's one of the most susceptible types of construction used today. The homeowner has a false sense of security. With time most slabs will crack. Termites can enter through tiny cracks in the concrete, over the edge of the slab, or easier yet, through openings around plumbing. Be sure the soil under your house is treated with the right type of chemical before pouring the slab and that the pest control operator uses the proper amount of the chemical.

WOOD SCREWS

The common types of wood screws have flat, oval, or round heads. The flathead screw is most commonly used if a flush surface is desired. Ovalhead and roundhead screws are used for appearance, and roundhead screws are used when countersinking is objectionable. Besides the head, the principal parts of a screw are the shank, thread, and core (fig. 7–4). Wood screws are usually made of steel or brass or other metals, alloys, or with specific finishes such as nickel, blued, chromium, or cadmium. They are classified according to material, type, finish, shape of head, and diameter of the shank or gage.

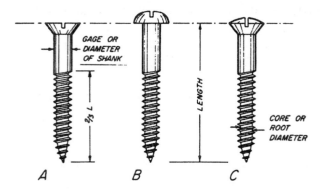

Figure 7–4.—Common types of wood screws: A, flathead; B, roundhead; and C, ovalhead.

Current trends in fastenings for wood also include tapping screws. Tapping screws have threads the full length of the shank and thus may have some advantage for certain specific uses.

Withdrawal Resistance

Experimental Loads

The resistance of wood screw shanks to withdrawal from the side grain of seasoned wood varies directly with the square of the specific gravity of the wood. Within limits, the withdrawal load varies directly with the depth of penetration of the threaded portion and the diameter of the screw, provided the screw does not fail in tension. The limiting length to cause screw failure decreases as the density of the wood increases. The longer lengths of standard screws are therefore superfluous in dense hardwoods.

The withdrawal resistance of type A tapping screws, commonly called sheet metal screws, is in general about 10 percent higher than for wood screws of comparable diameter and length of threaded portion. The ratio between the withdrawal resistance of tapping screws and wood screws is somewhat higher in the denser woods than in the lighter woods.

Ultimate test values for withdrawal loads of wood screws inserted into the side grain of seasoned wood may be expressed as:

$$p = 15,700G^2DL \qquad (7–7)$$

where p is the maximum withdrawal load in pounds, G is the specific gravity based on oven-dry weight and volume at 12 percent moisture content, D is the shank diameter of the screw in inches, and L is the length of penetration of the threaded part of the screw in inches.

This formula is applicable when screw lead holes in softwoods have a diameter of about 70 percent of the root diameter of the threads, and in hardwoods about 90 percent. The root diameter for most sizes of screws averages about two-thirds of the shank diameter.

The equation values are applicable to the following sizes of screws:

Screw length	Gage limits
Inches	
½	1 to 6
¾	2 to 11
1	3 to 12
1½	5 to 14
2	7 to 16
2½	9 to 18
3	12 to 20

For lengths and gages outside these limits, the actual values are likely to be less than the equation values. The withdrawal loads of screws inserted into the end grain of wood are somewhat erratic; but, when splitting is avoided, they should average 75 percent of the load sustained by screws inserted into the side grain.

Lubricating the surface of a screw is recommended to facilitate insertion, especially in the dense woods. It will have little effect on ultimate withdrawal resistance.

Allowable Loads

For allowable values, the practice has been to use one-sixth the ultimate load for longtime loading conditions. This also accounts for variability in test data. For normal duration of load, the allowable load may be increased 10 percent.

Glossary

A-frame A building design where the pitch of the roof is steep enough to essentially form an A and the roof serves as side walls.

Air-dried lumber Lumber that has been piled up in yards or sheds for any length of time. For the United States as a whole, the minimum moisture content of thoroughly air-dried lumber is 12 to 15 percent and the average is somewhat higher.

Anchor bolt A bolt to secure a wooden sill plate to concrete or a masonry floor or a foundation wall.

Apron The flat member of the inside trim of a window placed against the wall immediately beneath the stool.

Asbestos A silica-based mineral that was commonly used in many building materials. Asbestos fibers can cause lung cancer.

Asphalt Most native asphalt is a residue from evaporated petroleum. It is insoluble in water but soluble in gasoline and melts when heated. It is used widely as a waterproofing agent in the manufacture of many waterproof roof coverings, exterior wall coverings, flooring tile, and the like.

Attic ventilator A screened opening provided to ventilate an attic space. They are located in the soffit area as inlet ventilators and in the gable end or along the ridge as outlet ventilators. Attic ventilation can also be provided by means of power-driven fans. See Louver.

Backfill The replacement of excavated earth into a trench around and against an exterior basement wall.

Baluster A vertical member in a railing used on the edge of stairs, balconies, and porches.

Balustrade A railing held up by balusters.

Band joist A horizontal member that is nailed perpendicular to floor or ceiling joists and sits on the sill plate or top plate. See Joist.

Base or baseboard A board placed against the wall around a room next to the floor.

Base molding Molding used to trim the upper edge of the baseboard.

Base shoe Molding used next to the floor on the baseboard. Sometimes called a carpet strip.

Batten A narrow strip of wood used to cover joints or as decorative vertical members over plywood or wide boards.

Batter board One of a pair of horizontal boards nailed to posts set at the corners of an excavation, used to indicate the desired level. They are also used as fastenings for stretched strings to indicate outlines of foundation walls.

Bay window Any window space projecting outward from the walls of a building, either square or polygonal in plan.

Beam A structural member supporting a load applied transversely to the member.

Bedding A layer of mortar into which brick or stone is set.

Berm A raised area of earth such as earth pushed against a wall.

Blind-nailing Nailing in such a way that the nailheads are not visible on the face of the work. Blind nailing is usually done at the tongue of matched boards.

Blue stain See Sapstain.

Bolster A short horizontal wood or steel beam on a top of a column to support and decrease the span of beams or girders.

Boston ridge A method of applying shingles at the ridge or hips of a roof as a finish.

Brace An inclined piece of framing lumber applied to a wall or floor to stiffen the structure. Braces are often used temporarily on walls until the framing has been completed.

Brash Wood that breaks quickly and abruptly across the grain is said to be brash. Brashness is characteristic of even slightly decayed wood.

Breather paper A paper that lets water vapor pass through, often used on the outer face of walls to stop wind and rain but not trap water vapor.

Brick veneer A facing of brick laid against and fastened to the sheathing of a frame wall or tile wall construction.

Bridging Small wood or metal members inserted in a diagonal position between the floor joists at midspan to brace the joists.

Brown rot A type of decay common in softwoods that leaves the wood with an abnormal brown color as if it had been charred. Cross-grain cracking, collapse or crumbling, and abnormal shrinkage of the wood characterize advanced brown rot.

Built-up roof A roofing composed of three to five layers of asphalt felt laminated with coal tar, pitch, or asphalt. The top layer is covered with crushed slag or gravel. Generally used on flat or low-pitched roofs.

Butt joint The junction where the ends of two timbers or other members meet in a square-cut joint.

Cantilever A horizontal structural component that projects beyond its support, such as a second-story floor that extends out from the wall of the first floor.

Cant strip A triangular-shaped piece of lumber used at the junction of a flat deck and a wall to prevent cracking of the roofing that is built over it.

Cap The upper member of a column, pilaster, door cornice, or molding.

Carriage See Stringer.

Casement frame and sash A frame of wood or metal enclosing part or all of a sash that can be opened by means of hinges affixed to the vertical edge.

Casing Molding of various widths, forms, and thicknesses used to trim door and window openings at the jambs.

Caulk Used to fill or close a joint with a seal making it watertight and airtight. The material used to seal a joint.

Cement mortar A mixture of cement with sand and water used as a bonding agent between bricks or stones.

Chalking The slow release of individual particles from a painted surface by weathering.

Chalking string A string covered with chalk that, when stretched between two points and snapped, marks a straight line between the points.

Chamber The beveled edge of a board.

Checking (checks) Fissures that can appear with age in exterior paint coatings. Such fissures, at first superficial, may in time penetrate entirely through the coating. Also applies to narrow longitudinal openings that can appear on the surface of solid wood and veneer.

Check rails Also called meeting rails. The upper rail of the lower sash and the lower rail of the upper sash of a double-hung window. Meeting rails are made sufficiently thicker than the rest of the sash frame to close the opening between the two sashes. Check rails are usually beveled to ensure a tight fit between the two sashes.

Chimney connecter A stovepipe that is used to connect a stove to a masonry or prefabricated metal chimney.

Cleat A length of wood fixed to a surface, as a ramp, to give a firm foothold or to keep an object in place.

Clip A metal clip into which edges of adjacent plywood sheets are inserted to hold the edges in alignment.

Collar beam A horizontal member that is nominally 1 or 2 in. thick and connects opposite roof rafters at or near the ridge board. Collar beams stiffen the roof structure.

Column In architecture, a vertical supporting member, circular or rectangular in section, usually consisting of a base, shaft, and capital. In engineering, a vertical structural compression member that supports loads acting in the direction of its longitudinal axis.

Condensation Beads or films of water, or frost in cold weather, that accumulate on the inside of the exterior covering of a building when warm, moisture-laden air from the interior reaches a point where the air temperature no longer permits it to sustain the moisture it holds as vapor.

Conduction The transfer of heat through a material. Insulation is a poor conductor of heat, whereas metal is a good one.

Construction, frame A type of construction in which the structural parts are wood or depend upon a wood frame for support. In codes, if masonry veneer is applied to the exterior walls, the structure is still classified as frame construction.

Control joint A joint that penetrates only partially through a concrete slab or wall so that if cracking occurs, it will be on a straight line at that joint.

Convection A method of heat transfer by which lighter warm air rises and denser, cooler air moves downward.

Coped joint See Scribing.

Corner board A board used as trim or the external corner of a house or other frame structure, against which the ends of the siding are butted.

Corner brace A diagonal brace placed at the corner of a frame structure to stiffen and strengthen the wall.

Cornice Overhang of a pitched roof at the eave line, usually consisting of a fascia board, a soffit for a closed cornice, and appropriate moldings.

Cornice return The underside of the cornice at the corner of the roof where the walls meet the gable-end roof line. The cornice return serves as trim rather than as a structural element, providing a transition from the horizontal eave line to the sloped roof line of the gable.

Counterflashing A flashing usually used on chimneys at the roof line to cover shingle flashing and to prevent moisture from entering.

Course A continuous horizontal range of blocks, bricks, siding boards, or shingles.

Cove molding A molding with a concave face used as trim or to finish interior corners.

Crawl space A shallow space below the living quarters of a house without a basement, normally enclosed by the foundation wall.

Cricket See Saddle.

Crimp A crease formed in sheet metal for fastening purposes or to make the material less flexible.

Cripple stud A stud that extends from the floor to support a header.

Cross-grain cracking Severe cracking of paint, with excessive buildup, across the direction in which it was brushed on.

Crown molding A molding used on the cornice or wherever an interior angle is to be covered. If a molding has a concave face, it is called a cove molding.

Dead load The weight, expressed in mass per unit area, of elements that are part of the structure.

Deadman timber A large buried timber used as an anchor, as for anchoring a retaining wall.

Decay Disintegration of wood or another substance by fungal action, as opposed to insect damage.

Deck paint An enamel with a high degree of resistance to mechanical wear, designed for use on surfaces such as porch floors.

Density The mass of substance in a unit volume, such as pounds per cubic foot. When expressed in the metric system, it is numerically equal to the specific gravity of the same substance.

Dewpoint Temperature at which a vapor begins to deposit as a liquid. Applies especially to water in the atmosphere.

Dimension See Lumber, dimension.

Doorjamb The surrounding case into which and out of which a door closes and opens. It consists of two upright pieces, called side jambs, and a horizontal head jamb.

Dormer A roofed projection from a sloping roof, into which a dormer window is set. See Eye dormer and Shed dormer.

Downspout A pipe, usually of metal, that carries rainwater from roof gutters.

Dressed and matched See Tongue-and-groove.

Drip (a) A structural member of a cornice or other horizontal exterior-finish course that has a projection beyond the other parts for water runoff. (b) A groove in the underside of a sill or drip cap to cause water to run off on the outer edge.

Drip cap A molding placed on the exterior top side of a door or window frame to cause water to run off beyond the outside of the frame.

Dry rot An erroneous term used to describe well-decomposed, brown-rotted wood. The wood is dry upon current examination, but was wet when the decay occurred.

Drywall Interior covering material that is applied in large sheets or panels. The term has become basically synonymous with gypsum board.

Ducts Round or rectangular metal pipes for circulating warm air in a forced-air heating or air-conditioning system.

Earlywood (springwood) The band of light-colored and low-density wood formed in an annual ring at the beginning of a growing season.

Eave The lower margin of a roof projecting over the wall.

Edge-nailing Nailing into the edge of a board.

Efflorescence A white powdery substance formed on the surface of concrete or other masonry as the result of moisture movement through the material.

Enamel An opaque finish or paint that contains a relatively low proportion of pigment and dries to a sheen or luster.

End-nailing Nailing into the end of a board, which results in very poor withdrawal resistance.

Equilibrium moisture content The moisture content of wood when it has been allowed to equalize itself in relation to the surrounding temperature and relative humidity.

Expansion joint A bituminous fiber strip used to separate blocks or units of concrete to prevent cracking caused by expansion that is a result of temperature changes. Also used on concrete slabs.

Eye dormer A dormer that has a gable roof.

Face-nailing Nailing perpendicular to the initial surface being penetrated. Also termed direct nailing.

Fascia A flat board, band, or face used by itself or, more often, in combination with moldings and generally located at the outer face of the cornice.

Fascia backer The main structural support member to which the fascia is nailed.

Felt See Saturated felt.

Fiberboard A reconstituted board material made from mechanical pulp. The board may be impregnated with asphalt and used as a wall sheathing.

Fiberglass A common type of insulation composed of very fine inorganic fibers made from rock, slag, or glass with other materials added to enhance its useability.

Filler In wood finishing, a paste or liquid with or without coloring pigment that is applied to open-grained hard-woods such as oak to fill the pores and smooth the surface.

Filler A heavily pigmented preparation used for filling and leveling off the pores in open-pored woods.

Fire stop A solid, tight closure of a concealed space, placed to prevent the spread of fire and smoke. In a frame wall, this usually consists of 2 by 4 cross-blocking between studs.

Flagstone (flagging or flags) Flat stones, 1 to 4 in. thick, used for rustic walks, steps, and floors.

Flakeboard A panel material made of specially produced flakes that are compressed and bonded together with phenolic resin. Popular types include waferboard and oriented strandboard (OSB). Structural flakeboards are used for many of the same applications as plywood.

Flashing Strips of metal or asphalt composition material that are used to exclude moisture where the roof intersects outside walls, chimneys, vents, and roof valleys.

Flat paint An interior paint that contains a high proportion of pigment and dries to a flat or lusterless finish.

Flue The space or passage in a chimney through which smoke, gas, or fumes ascend. Each such passage is called a flue, which together with any others and the surrounding masonry make up the chimney.

Flue lining Fire clay or terra cotta pipe, round or square, usually made in all ordinary flue sizes and in 2-ft lengths, used for the inner lining of chimneys with the brick or masonry work around the outside. Flue lining in chimneys runs from about a foot below the flue connection to the top of the chimney.

Fly rafter End rafters of the roof overhang supported by roof sheathing and lookouts.

Footnotes A concrete section in a rectangular form, wider than the bottom of the foundation wall or pier it supports. With a pressure-treated wood foundation, a gravel footing may be used in place of concrete.

Formwork A temporary mold for giving a desired shape to poured concrete.

Foundation The supporting portion of a structure below the first-floor construction, or below grade.

Framing, balloon A system of framing in which all exterior studs extend in one piece from the sill plate to the roof plate.

Framing, ladder Framing for the roof overhang at a gable. Cross pieces are used similarly to a ladder to support the overhang.

Framing, platform A system of framing in which floor joists of each story rest on the top plates of the story below or on the foundation sill for the first story and the bearing walls and partitions rest on the subfloor of each story.

Frieze A horizontal member connecting the top of the siding with the soffit of the cornice.

Frost line The depth of frost penetration in soil. This depth varies in different parts of the country.

Fungi, wood Microscopic plants that live in damp wood and cause mold, stain, and decay.

Fungicide A chemical that is poisonous to fungi.

Furring strips Strips of wood or metal applied to a wall or other surface to even it and to serve as a fastening base for finish material.

Gable The portion of the roof above the eave line of a double-sloped roof.

Gable dormer A vertical window in a projection built out from a sloping roof. The dormer itself has a hip roof.

Gable end An end wall having a gable.

Gambrel A roof that slopes steeply at the edge of a building but changes to a shallower slope across the center of the building. This allows the attic to be used as a second story.

Glazing putty The initially soft workable material that is used to seal between the glass in a window and the wood frame that holds it.

Glue line, exterior Waterproof glue at the interface of two veneers of plywood.

Girder A large or principal beam of wood or steel used to support loads at points along its length.

Grade The ground level around a building. The natural grade is the original level. Finished grade is the level after the building is complete and final grading is done.

Grain The direction, size, arrangement, appearance, or quality of the fibers in wood.

Grain, edge (vertical) Edge-grained lumber has been sawed parallel to the pith of the log and approximately at right angles to the growth rings; that is, the rings form an angle of 45° or more with the wide surface of the piece.

Grain, flat Flat-grained lumber has been sawed parallel to the pith of the log and approximately tangential to the growth rings; that is, the rings form an angle of less than 45° with the surface of the piece.

Granulated cork A loose-fill type of insulation.

Ground fault circuit interrupter (GFCI) A device that cuts the power to an outlet if current from that outlet is grounded by human contact.

221

Grout Mortar that will flow into the joints and cavities of masonry work and fill them solidly.

Gusset A flat wood, plywood, or similar type member used to provide a connection at intersections of wood members. Most commonly used in joints of wood trusses.

Gutter or eave trough A shallow channel or conduit of metal or vinyl set below and along the eaves of a house to catch and carry off rainwater from the roof.

Gypsum board See Drywall.

HVAC Common abbreviation for heating, ventilation, and air-conditioning equipment.

H-clip A metal clip into which edges of adjacent plywood sheets are inserted to hold them in alignment.

Hardboard A relatively heavy type of fiberboard made from mechanical pulps. It may be used in 4- by 8-ft sheets for leveling floors before applying sheet vinyl, etc., or coated and used for wall paneling. In a tempered form, it is used in 4- by 8-ft sheets for exterior siding or manufactured as a substitute for bevel drop siding.

Header (a) A beam placed perpendicular to joists, to which joists are nailed in framing for chimneys, stairways, or other openings. (b) A wood lintel.

Hearth The inner or outer floor of a fireplace, usually made of brick, tile, or stone.

Heartwood The wood extending from the pith to the sapwood, the cells of which no longer participate in the life process of the tree.

High-density overlay (HDO) A special hard, phenolic-resin-impregnated fiber surface that can be applied to plywood. It is commonly used where a skid-resistant finish is needed.

Hip The external angle formed by the meeting of two sloping sides of a roof.

Hip roof A roof that rises by inclined planes from all four sides of a building.

Hopper window A window that is hinged at the bottom to swing inward.

Humidifier A device that increases the humidity within a room or a house by discharging water vapor. Humidifiers may consist of individual room-size units or larger units attached to the heating unit to condition the entire house.

Humidity The amount of water vapor in the atmosphere, expressed as a percentage of the maximum quantity that the atmosphere could hold at a given temperature. The amount of water vapor that can be held in the atmosphere increases with the temperature.

Hyphae Microscopic thread-like substance of a fungus that spreads through wood and causes decay.

I-beam A steel beam with a cross section resembling the letter I. I-beams are used for long spans as basement beams or over wide wall openings, such as a double garage door, when wall and roof loads are imposed on the opening.

Ice dams When snow melts on the heated portion of the roof or from the sun, the water can flow to the eave, refreeze, and form a ridge of ice. As additional moisture moves down the roof, it may back up under the shingles and leak into the attic or side walls.

Incipient wood decay The early stage of wood decay, which is difficult to detect without the use of a microscope or culturing techniques.

Insulation board, rigid A building board commonly used as sheathing and made of polystyrene, polyurethane, or other plastics; mineral fiber; or coarse wood or cane fiber impregnated with asphalt or given other treatment to provide a water-resistant product. It can be obtained in various sizes, thicknesses, and densities.

Insulation, thermal Any material high in resistance to heat transmission that, when placed in the walls, ceiling, or floors of a structure, reduces the rate of heat flow.

Isolation joint A joint in which two incompatible materials are isolated from each other to prevent chemical action between the two.

Jack rafter A rafter that spans the distance from the wall plate to a hip or from a valley to a ridge.

Jack stud A short stud that does not extend from floor to ceiling; for example, a stud that extends from the floor to a window.

Jamb The surrounding of a doorway, window, or other opening. It consists of two upright pieces called side jambs and a horizontal head jamb.

Joint The space between the adjacent surfaces of two members or components that are held together by nails, glue, cement, mortar, or other means. See Control joint, Coped joint, Expansion joint, and Isolation joint.

Joint cement A powder that is usually mixed with water and used to treat joints in gypsum board. Joint cement, often called spackle, can be purchased in a ready-mixed form.

Joist One of a series of parallel beams, usually 2 in. thick, used to support floor and ceiling loads, and supported in turn by larger beams, girders, or bearing walls. See Band joist, Header, Tail joist, and Trimmer.

Joist hanger A metal bracket that is used to support the ends of a joist.

Kerf A cut or incision made by a saw in a piece of wood.

Keyways A tongue-and-groove type connection where perpendicular planes of concrete meet to prevent relative movement between the two components.

Kiln-dried lumber Lumber that has been dried by controlled heat and humidity, in ovens or kilns, to specified ranges of moisture content. See Air-dried lumber and Lumber, moisture content.

Knee wall A short wall extending from the floor to the roof in the second story of a 1-1/2-story house.

Lacquer A clear finish noted for its quick drying properties that is often used on furniture, cabinets, and sometimes woodwork.

Landing A platform between flights of stairs or at the end of a flight of stairs.

Latewood (summerwood) The band of dark, dense, and hard wood formed in an annual ring towards the end of the growing season in species such as southern pine and Douglas-fir.

Lath Thin, narrow strips of wood nailed horizontally to the interior side of stud walls to which plaster is applied.

Lay up To place materials together in the relative positions they will have in the finished building.

Ledger plate A strip of lumber nailed along the bottom of the side of a girder, on which joists rest.

Let-in brace A board nominally 1 in. thick installed diagonally into notched studs.

Light Space in a window sash for a single pane of glass. Also, a pane of glass.

Lintel A horizontal structural member that supports the load over an opening such as a door or window. Also called a header.

Live load The load, expressed in mass per unit area, of people, furniture, snow, and so forth that is in addition to the weight of the structure itself.

Lookout A short wood bracket or cantilever to support an overhand portion of a roof, usually concealed from view by a soffit. See Fly rafters.

Louver An opening with a series of horizontal slats arranged to permit ventilation but to exclude rain, sunlight, or vision. See Attic ventilator.

Lumber, boards Lumber less than 2 in. thick and 2 in. or more wide.

Lumber, dimension Lumber from 2 in. thick to, but not including, 5 in. thick and 2 in. or more wide. Includes joists, rafters, studs, planks, and small timbers.

Lumber, dressed size The dimension of lumber after shrinking from green dimension and after machining to size or pattern.

Lumber, matched See Tongue and groove.

Lumber, moisture content The weight of water contained in wood, expressed as a percentage of the oven-dry weight of the wood. See Air-dried lumber and Kiln-dried lumber.

Lumber, pressure-treated Lumber that has had a preservative chemical forced into the wood under pressure to resist decay and insect attack.

Lumber, shiplap Lumber that has been milled along the edge to make a close rabbeted or lapped joint.

Lumber, timbers Lumber 5 in. or more in the least dimension. Includes beams, stringers, posts, caps, sills, girders, and purlins.

Mansard A type of roof that slopes very steeply around the perimeter of the building to full wall height, providing space for a complete story. The center portion of the roof is either flat or very shallow-sloped.

Mantel The shelf above a fireplace. Also used to refer to the trim around both top and sides of a fireplace opening.

Masonry Stone, brick, concrete, hollow-tile, concrete block, gypsum block, or other similar building units or materials or a combination of the same, bonded together with mortar to form a wall, pier, buttress, or similar element.

Mastic A pasty material used as a cement for setting tile or as a protective coating for thermal insulation or waterproofing.

Medium-density overlay (MDO) A resin-impregnated paper coating used on plywood. It provides an excellent base for painting.

Mildew A term that applies both to the fungus (a type of microscopic plant life) and to its stain and degenerative effects on paint or wood.

Millwork Building materials made of finished wood and manufactured in millwork plants and planing mills. It includes such items as inside and outside window and door frames, blinds, porch work, mantels, panel work, stairways, molding, and interior trim. The term does not include flooring or siding.

Mineral fiber See Fiberglass.

Miter joint The joint of two pieces at an angle that is half the joining angle. For example, the miter joint at the side and head casing at a door opening is made at a 45° angle.

Moisture content of wood See Lumber, moisture content.

Molding A wood strip with a curved or projecting surface used for decorative purposes.

Molds Generally a green, black, or occasionally orange or other light discoloration on the surface of lumber. Molds can generally be planed or even brushed off. Strength, other than toughness, is not affected.

Mortar See Cement mortar.

Mortise A slot cut into a board, plank, or timber, usually edgewise, to receive a tenon of another board, plank, or timber to form a joint.

Mullion A vertical bar or divider in the frame between windows, doors, or other openings.

Muntin A small member that divides the glass or openings of sashes or doors.

Natural finish A transparent finish that does not seriously alter the original color or obscure the grain of the natural wood. Natural finishes are usually provided by sealers, oils, varnishes, water-repellent preservatives, and other similar materials.

Newel A post to which the end of a stair railing or balustrade is fastened. Also, any post to which a railing or balustrade is fastened.

Nominal (wood dimension) The approximate size of a sawn wood section before it is planed.

Nonload-bearing wall A wall supporting no load other than its own weight.

Nosing The projecting edge of a molding or drip. Usually applied to the projecting molding on the edge of a stair tread.

Notch A crosswise rabbet at the end of a board.

On center (O.C.) The measurement of spacing for elements such as studs, rafters, and joists, from the center of one member to the center of the next.

Opaque finish A finish such as paint that covers the wood grain.

Oriented strandboard (OSB) A type of structural flakeboard composed of layers, with each layer consisting of compressed strand-like wood particles in one direction, and with layers oriented at right angles to each other. The layers are bonded together with a phenolic resin.

Panel (a) A thin flat piece of wood, plywood, or similar material, framed by stiles and rails as in a door or fitted in grooves to thicker material with molded edges for decorative wall treatment. (b) A sheet of plywood, fiberboard, structural flakeboard, or similar material.

Paper, building A general term for papers, felts, and similar sheet materials used in construction.

Parquet A floor with inlaid design. For wood flooring, it is often laid in blocks with boards at angles to each other to form patterns.

Particleboard Panels composed of small wood particles usually arranged in layers by particle size without a particular orientation and bonded together with a phenolic resin. Some particleboards are structurally rated. See also Structural flakeboard.

Parting strip A small wood piece used in the side and head jambs of double-hung windows to separate the upper and lower sash.

Partition A wall that subdivides spaces within any story of a building.

Penny As applied to nails, it originally indicated the price per hundred. The term serves as a measure of nail length and is signified by the letter "d."

Perlite A loose-fill type of aggregate formed from volcanic ash. It may be used to fill hollow-core masonry walls where it acts as insulation.

Perm A measure of water vapor movement through a material (grains per area per hour per unit difference in vapor pressure).

Pick test A general test used to determine if wood is brash or if appreciable decay has occurred. A sharp pointed tool is pushed into the wood and then pried out breaking the surface of the wood in the process. The break is classified as brash or splintery. A splintery break indicates sound wood.

Pier A column of masonry, usually rectangular in horizontal cross section, used to support other structural members.

Pier and beam construction A type of construction where masonry columns (piers) are used to support horizontal beams on which floor joists are placed to form a crawl-space-type house.

Pigment A powdered solid in suitable degree of subdivision for use in paint or enamel.

Pilaster A projection from a wall forming a column to support the end of a beam framing into the wall.

Picocurie A measure of radioactivity. The metric equivalent is becquerel.

Pitch The measure of the steepness of the slope of a roof, expressed as the ratio of the rise of the slope over a corresponding horizontal distance. Roof slope is expressed in the inches of rise per foot of run, such as 4 in 12.

Pitch board A template used for marking the rise and run on a stair carriage.

Plate Sill plate, a horizontal member anchored to a masonry wall; sole plate, a bottom horizontal member of a frame wall; top plate, top horizontal member of a frame wall supporting ceiling joists, rafters, or other members.

Plenum The space in which air is contained under pressure slightly greater than atmospheric pressure. In a house, it is used to distribute heated or cooled air.

Plugged A method by which defects in plywood are cut out and another piece of the same size is inserted and glued in place. The face plys are commonly plugged to increase the grade. The inner plys may be treated the same way for certain specialty grades.

Plumb Exactly vertical.

Plumb bob A metal weight or bob used at the end of a plumb line. A plumb line is used to determine verticality or if something is plumb.

Ply A term denoting one thickness of any material used for the building up of several layers, such as roofing felt, veneer in plywood, or layers in built-up materials.

Plywood A piece of wood made of three or more layers of veneer joined with glue and usually laid with the grain of adjoining plies at right angles.

Pocket rots Spindle shaped, pointed, white or brown pockets or cavities parallel to the grain and separated by sound wood. These rots occur in the living tree and do not further damage lumber after it is cut.

Polystyrene A thermoplastic-type insulation available as expanded plastic beads (bead boards) and extruded foam. The R-value may range from 3.6 to 5.3 per inch.

Polyurethane A type of insulation that is available in rigid board form, or it may be foamed into place. The material has an exceptionally high R-value of about 6.3.

225

Post-and-beam roof A roof consisting of thick planks spanning between beams that are supported on posts. This construction has no attic or air space between the ceiling and roof.

Powder-post beetles Beetles that convert the inner portion of infected wood to a powdery or pelleted mass.

Primer The first coat of paint in a paint job that consists of two or more coats; also the paint used for such a first coat.

Pressure-treated lumber See Lumber, pressure-treated.

Pressure-treated wood Wood that has been treated with a wood preservative under pressure. When properly treated, the wood resists decay and insect attack.

Purlin A horizontal timber supporting the common rafters in roofs.

Quality mark A quality control ink mark on some pressure-treated lumber that indicates the treating company, the type and amount of treatment, and other information. The American Wood Preservers Bureau is responsible for administering the program.

Quarter round A small molding that has the cross section of a quarter circle.

Quartersawn Another term for edge grain.

R-value The ability of a material to retard heat conduction. The higher the R-value, the higher the insulating value.

Rabbet A rectangular longitudinal groove cut in the corner edge of a board or plank.

Radiant heat The effect that is felt when sitting by a warm stove. Radiation is the process by which heat is transferred from the stove to the object.

Radon A colorless, odorless, radioactive gas that naturally occurs virtually everywhere in the earth and, when concentrated or confined, may cause lung cancer.

Rafter One of a series of structural members of a roof designed to support roof loads. The rafters of a flat roof are sometimes called roof joists. See Fly rafter; Jack rafter; Rafter, hip; and Rafter, valley.

Rafter, hip A rafter that forms the intersection of an external roof angle.

Rafter, valley A rafter that forms the intersection of an internal roof angle. A valley rafter is normally made of double 2-in.-thick members.

Rail (a) A cross-member of a panel door or sash. (b) The upper or lower member of a balustrade or staircase extending from one vertical support, such as a post, to another.

Rake Trim members that run parallel to the roof slope and form the finish between the wall and a gable roof extension.

Reflective insulation Sheet material with one or both surfaces of comparatively low heat emissivity, such as aluminum foil. When it is used in building construction, the surfaces face air spaces, reducing the radiation across these spaces.

Register A device for controlling the flow of warmed or cooled air through an opening.

Reinforcing Steel rods or metal fabric placed in concrete slabs, beams, or columns to increase their strength.

Resorinol An adhesive that is high in both wet and dry strength and is resistant to high temperatures. It is used for gluing lumber or assembly joints that must withstand severe conditions.

Reverse board and batten (plywood finish) Siding in which narrow battens are nailed vertically to wall framing and wider boards are nailed over these so that the edges of boards lap the battens. A slight space is left between adjacent boards. This pattern is simulated with plywood by cutting wide vertical grooves in the face ply at uniform spacing.

Ridge line The horizontal line at the junction of the top edges of two sloping roof surfaces.

Ridge board The board placed on edge at the ridge of the roof, into which the upper ends of the rafters are fastened.

Ring shank nail A nail with ridges forming rings around the shank to provide better withdrawal resistance.

Rise In stairs, the vertical height of a step or flight of stairs.

Riser Each vertical board closing the spaces between the treads of stairways.

Rock wool See Fiberglass.

Roll roofing Roofing material composed of fiber and saturated with asphalt and supplied in 36-in.-wide rolls with 108 ft^2 of material. Weights are generally 45 to 90 lb per roll.

Roof, built-up See Built-up roof.

Roof, sheathing The boards or sheet material fastened to the roof rafters, on which shingles or other roof covering is laid.

Roof, valley See Valley.

Roofing paper See Roll roofing.

Run In stairs, the net front-to-back width of a step or the horizontal distance covered by a flight of stairs.

Saddle Two sloping surfaces meeting in a horizontal ridge, used between the back side of a chimney, or other vertical surface, and a sloping roof. Saddles are also called crickets.

Sapstain (also **Blue stain**) A blue, black, gray, or brown darkening of sapwood caused by large masses of dark fungal hyphae deeply penetrating the cells. Sapstains do not seriously decrease wood strength except for toughness.

Sapwood The outer zone of wood in a tree, next to the bark. In a living tree, sapwood contains some living cells (heartwood contains none), as well as dead and dying cells. In most species, it is lighter in color than heartwood. In all species, it lacks resistance to decay.

Sash A frame containing one or more lights of glass.

Saturated felt Felt impregnated with tar or asphalt.

Sawkerf See Kerf.

Scab. A short length of board nailed over the joint of two boards butted end-to-end to transfer tensile stresses between the two boards.

Scaling Losing the smooth surface of concrete as part of the surface comes off in flakes or scales.

Screw shank nail A nail with spiral ridges around the shank to provide better withdrawal resistance.

Scribing Fitting woodwork to an irregular surface. With moldings, scribing means cutting the end of one piece to fit the molded face of the other at an interior angle, in place of a miter joint.

Scuttle hole An opening in the ceiling to provide access to the attic. It is closed by a panel when not in use.

Sealant See Caulk.

Sealer A finishing material, either clear or pigmented, that is usually applied directly over uncoated wood to seal the surface.

Seam, standing A joint between two adjacent sheets of metal roofing in which the edges are bent up to prevent leaking and the joint between the raised edges is covered.

Seasoning Removing moisture from green wood to improve its serviceability.

Semigloss paint or enamel A paint or enamel made with a slight insufficiency of nonvolatile vehicle so that its coating, when dry, has some luster but is not very glossy.

Semitransparent penetrating stains Moderately pigmented water repellents or water-repellant preservatives that can color a wood surface. The finish penetrates the wood surface and does not form a surface film.

Shake A thick handsplit shingle, resawed to form two shakes; usually edge-grained.

Sheathing The covering used over the joists, studs, or rafters of a structure.

Sheathing paper A building material, generally paper or felt, used in wall and roof construction as protection against the passage of air and water.

Shed dormer A dormer that has a roof sloping only one direction at a much shallower slope than the main roof of the house.

Sheet metal work All components of a house employing sheet metal, such as ducts, flashing, gutters, and downspouts.

Shellac A transparent coating made by dissolving lac, a resinous secretion of the lac bug (a scale insect that thrives in tropical countries, especially India), in alcohol.

Shim A thin wedge of wood for driving into crevices to bring parts into alignment.

Shingles Roof covering of asphalt, fiberglass, wood, tile, slate, or other material or combinations of materials such as asphalt and felt, cut to stock lengths, widths, and thicknesses.

Shingle, siding Various types of shingles used over sheathing for exterior sidewall covering.

Shiplap See Lumber, shiplap.

Shutter A lightweight louvered flush wood or nonwood frame in the form of a door, located at each side of a window. Some are made to close over the window for protection; others are fastened to the wall for decorative purposes.

Siding bevel (lap siding) Wedge-shaped boards used as horizontal siding in a lapped pattern. Bevel siding varies in butt thickness from 1/2 to 3/4 in. and is available in widths up to 12 in. Normally used over some type of sheathing.

Siding, drop Siding that is usually 3/4 in. thick and 6 or 8 in. wide, with tongue-and-groove or shiplap edges. Often used as siding without sheathing in secondary buildings.

Sill (a) The lowest member of the frame of a structure, resting on the foundation and supporting the floor joists or the uprights of the wall. (b) The member forming the lower side of an opening such as a door sill or window sill.

Slab A concrete floor poured on the ground.

Sleeper A wood member embedded in or resting directly on concrete, as in a floor, that serves to support and to fasten subfloor or flooring.

Slip tongue A spline used to connect two adjacent boards that have grooves facing each other.

Soffit The underside of an overhanging cornice.

Soil cover (ground cover) A light covering of plastic film, roll roofing, or similar material, used over the soil in crawl spaces of buildings to minimize movement of moisture from the soil into the crawl space.

Soil stack A general term for the vertical main of a system of soil, waste, or vent piping.

Sole or sole plate See Plate.

Solid color stains These opaque finishes come in a wide range of colors and are essentially thin paints. They are also called hiding, heavy-bodied, opaque, or blocking stains.

Spackle See Joint cement.

Spalling Chips or splinters breaking loose from the surface of concrete because of moisture moving through from the reverse side.

Span The distance between structural supports.

Splash block A small masonry block laid with the top close to the ground surface, to receive roof drainage from downspouts and carry it away from the building.

Spline A long, narrow, thin strip of wood or metal often inserted into the edges of adjacent boards to form tight joints.

Spores Spores are produced when a fungus fruits. They are comparable to seeds in that they float everywhere through the air, and when they contact wet wood, germination occurs and decay can follow.

Square A unit of measure, usually applied to roofing material, that denotes a sufficient quantity to cover 100 ft^2 of surface.

Stair carriage Supporting member for stair treads. Usually a 2-in. plank notched to receive the tread; sometimes called a rough horse or stringer.

Stair landing See Landing.

Stair rise See Rise.

Sound Transmission Class (STC) A numerical measure of the ability of a material or assembly to resist the passage of sound. Materials with higher STC numbers have greater resistance to sound transmission.

Stile An upright framing member in a panel door.

Stool A flat molding fitted over the window sill between jambs and contacting the bottom rail of the lower sash.

Stop, trim The trim member on the jambs of an opening that a door or window closes against.

Stop, gravel A raised ridge of metal at the edge of a tar and gravel roof that keeps the gravel from falling off the roof.

Strip flooring Thin tongue-and-groove lumber.

String, stringer A timber or other support for cross members in floors or ceilings. In stairs, the stringer or stair carriage supports the stair treads.

Structural flakeboard A panel material made of specially produced flakes that are compressed and bonded together with phenolic resin. Popular types include waferboard and oriented strandboard (OSB). Structural flakeboards are used for many of the same applications as plywood.

Stucco A plaster for exterior use, made with Portland cement as its base.

Stud One of a series of slender wood or metal vertical structural members placed as supporting elements in walls and partitions. (Plural: studs or studding.)

Studwall A wall consisting of spaced vertical structural members with thin facing material applied to each side.

Subfloor Boards or plywood laid on joists, over which a finish floor is to be laid.

Tack cloth A sticky cloth used to remove dust before finishing wood.

Tail joist A relatively short joist or beam supported by a wall at one end and by a header at the other.

Tenon A projection at the end of a board, plank, or timber for insertion into a mortise.

Termite guards A shield, usually of noncorrodible metal, placed in or on a foundation wall or other mass of masonry or around pipes, to prevent passage of termites.

Terne (roof) Sheet iron or steel coated with an alloy of lead and tin used on roofs.

Thimble The section of a vitreous clay flue that passes through a wall.

Thinner A liquid that evaporates readily and is used to thin the consistency of finishes without altering the relative volumes of pigments and nonvolatile vehicles.

Threshold A strip of wood or metal with beveled edges, used over the finish floor and the sill of exterior doors.

Tieback member A timber oriented perpendicularly to a retaining wall that ties the wall to a deadman timber buried behind the wall.

Toe-nailing Driving a nail at a slant with the initial surface so it will penetrate a second member.

Tongue-and-groove Boards or planks machined in such a manner that there is a groove on one edge and a corresponding projection (tongue) on the other edge, so that a number of such boards or planks can be fitted together. Dressed and matched is an alternative term.

Tread The horizontal board in a stairway on which the foot is placed.

Trim The finish materials in a building, such as molding applied around openings (window trim, door trim) or at the floor and ceiling of rooms (baseboard, cornice, and other moldings).

Trimmer A beam or joist to which a header is nailed in framing a chimney, stairway, or other opening.

Truss A framed or jointed structure that is composed of triangular elements and designed to act as a beam of long span; each member is usually subjected to longitudinal stress only, either tension or compression.

Truss plate A heavy-gauge, pronged metal plate that is pressed into the sides of a wood truss at the point where two more members are to be joined together.

Undercoat A coating applied prior to the finishing or top coats of a paint job. When it is the first of two or more coats, it is synonymous with the priming coat.

Underlayment A material placed under flexible flooring materials such as carpet, vinyl tile, or linoleum, to provide a smooth base over which to lay such materials.

Valley The internal angle formed by the junction of two sloping sides of a roof.

Vapor retarder Material used to retard the movement of water vapor into walls. Vapor retarders are applied over the warm side of exposed walls or as a part of batt or blanket insulations. They usually have perm values of less than 1.0. Vapor retarders were initially called vapor barriers.

Varnish A thickened preparation of drying oil or drying oil and resin, suitable for spreading on surfaces to form continuous, transparent coatings, or for mixing with pigments to make enamels.

Vehicle The liquid portion of a finishing material, consisting of the binder (nonvolatile) and volatile thinners.

Vent A pipe or duct, or a screened or louvered opening that provides an inlet or outlet for the flow of air. Common types of roof vents include ridge vents, soffit vents, and gable end vents.

Vermiculite A loose-fill-type insulation with an R-value of 2.3 per in. It is made by the expansion of mica granules at high temperatures and is highly water-absorbent.

Waferboard A type of structural flakeboard made of compressed, wafer-like wood particles or flakes bonded together with a phenolic resin. The flakes may vary in size and thickness and may be either randomly or directionally oriented.

Wall plate The cover over an electrical outlet or switch on the wall.

Wane Bark, or lack of wood from any cause, on the edge or corner of a piece of wood. Hence, waney.

Water-repellent preservative A liquid designed to penetrate wood to make it water resistant and give moderate preservative protection. It is used for millwork such as sash and frames and is usually applied by dipping.

Water-repellent preservative finish These finishes contain a fungicide, a small amount of wax as a water repellent, a resin or drying oil, and a solvent such as turpentine or mineral spirits. The material is used as a natural finish and as a pretreatment before priming wood for painting.

Water repellents This finish is the same as a water-repellent preservative without the preservative. The solution helps to stabilize wood and thus prevents paint peeling.

Weatherstripping Strips of thin metal or other material that prevent infiltration of air and moisture around windows and doors. Compression weatherstripping on single- and double-hung windows also holds such windows in place in any position.

Web The thin center portion of a beam that connects the wider top and bottom flanges.

Wedges Triangular-shaped pieces of wood that can be driven into gaps between rough framing and finished items, such as window and door frames, to provide a solid backing for these items.

Weep pipe A pipe installed in a foundation wall to allow water to collect in a concentrated area.

Whaler A large structural member placed horizontally against foundation forms to which braces are temporarily attached to prevent forms from moving horizontally under the pressure of concrete.

White rot A type of decay common in hardwoods. Initially, the wood turns an off-white. Black zone lines may develop, and a white fibrous mass may result.

Withe A vertical layer of bricks, one brick thick.

Wood extractives Dark, water-soluble materials present in the heartwood of many species. These materials may leach from the wood and discolor paint.